Study Guide

T0331929

Physics Unit 1
for CAPE®

Terry David
Joyce Crichlow
Dwight de Freitas
Carlos Hunte

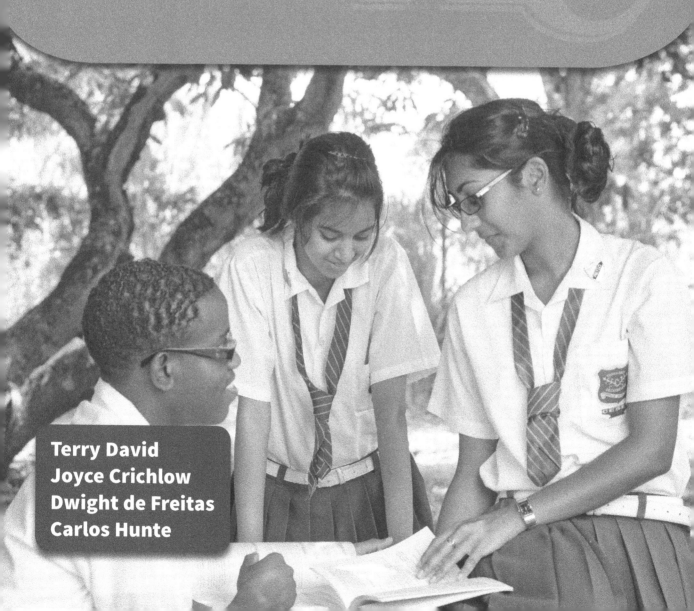

Great Clarendon Street, Oxford, OX2 6DP, United Kingdom

Oxford University Press is a department of the University of Oxford.
It furthers the University's objective of excellence in research, scholarship,
and education by publishing worldwide. Oxford is a registered trade mark of
Oxford University Press in the UK and in certain other countries

First published by Nelson Thornes Ltd in 2012
This edition published by Oxford University Press in 2014

British Library Cataloguing in Publication Data
Data available

978-1-4085-1761-1

14

Printed in Great Britain by Ashford Colour Ltd., Gosport

Acknowledgements

Cover photograph: Mark Lyndersay, Lyndersay Digital, Trinidad
www.lyndersaydigital.com
Illustrations: GreenGate Publishing Services, Tonbridge, Kent
Page make-up: GreenGate Publishing Services, Tonbridge, Kent

Although we have made every effort to trace and contact all
copyright holders before publication this has not been possible in all
cases. If notified, the publisher will rectify any errors or omissions at
the earliest opportunity.

Links to third party websites are provided by Oxford in good faith
and for information only. Oxford disclaims any responsibility for
the materials contained in any third party website referenced in
this work.
The manufacturer's authorised representative in the EU for
product safety is Oxford University Press España S.A. of El Parque
Empresarial San Fernando de Henares, Avenida de Castilla, 2 –
28830 Madrid (www.oup.es/en or product.safety@oup.com).
OUP España S.A. also acts as importer into Spain of products
made by the manufacturer.

Contents

Contents

Introduction

This Study Guide has been developed exclusively with the Caribbean Examinations Council (CXC®) to be used as an additional resource by candidates, both in and out of school, following the Caribbean Advanced Proficiency Examination (CAPE®) programme.

It has been prepared by a team with expertise in the CAPE® syllabus, teaching and examination. The contents are designed to support learning by providing tools to help you achieve your best in CAPE® Physics and the features included make it easier for you to master the key concepts and requirements of the syllabus. *Do remember to refer to your syllabus for full guidance on the course requirements and examination format!*

Inside this Study Guide is an interactive CD which includes electronic activities to assist you in developing good examination techniques:

- **On Your Marks** activities provide sample examination-style short answer and essay type questions, with example candidate answers and feedback from an examiner to show where answers could be improved. These activities will build your understanding, skill level and confidence in answering examination questions.

- **Test Yourself** activities are specifically designed to provide experience of multiple-choice examination questions and helpful feedback will refer you to sections inside the study guide so that you can revise problem areas.

- **Answers** are included on the CD for multiple-choice questions and questions that require calculations, so that you can check your own work as you proceed.

This unique combination of focused syllabus content and interactive examination practice will provide you with invaluable support to help you reach your full potential in CAPE® Physics.

1.1 SI quantities and units

Learning outcomes

On completion of this section, you should be able to:

- recall the SI base quantities and units
- determine the units of derived quantities
- recall commonly used prefixes
- convert units
- define the mole and recall the Avogadro's constant.

Table 1.1.1 SI base quantities and units

Physical quantity	Symbol	Unit
Mass	m	kilogram (kg)
Length	l	metre (m)
Time	t	second (s)
Temperature	T	kelvin (K)
Electric current	I	ampere (A)
Amount of substance	n	mole (mol)
Luminous intensity	I_v	candela (cd)

Physical quantities

A **physical quantity** is the property of an object or a phenomenon that can be measured with an instrument. It is typically expressed as the product of a numerical magnitude and a unit. For example, suppose a student measures his mass and records it as 55 kg. The numerical magnitude in this case is 55. The unit of mass is the kilogram (kg).

SI base quantities and units

Scientists worldwide have agreed on a common system of units known as Le Système Internationale d'Unites (The International System of Units) or **SI units**. All scientific measurements are made using these units. In this system there are seven **base units** which have been defined in such a way that they can be easily reproduced (Table 1.1.1). These units represent the standard size of a particular physical quantity.

Derived quantities and units

Physical quantities other than the base quantities are known as **derived quantities** (Table 1.1.2). A derived quantity is derived from a combination of base quantities. The corresponding unit is derived from the relationship between the base quantities. Speed is a derived quantity. It is defined by the following equation:

$v = \dfrac{s}{t}$ where v is speed, s is distance travelled and t is the time taken.

From the definition, distance is a base quantity with unit m (metre). Time is also a base quantity with unit s (second).

$$v = \frac{s}{t} = \frac{m}{s} = m\,s^{-1}$$

The SI unit for speed is therefore $m\,s^{-1}$.

Table 1.1.2 Derived quantities and derived units

Derived quantity	Relationship	Derived unit in base units	Name of unit
Area (A)	length × length	m^2	–
Volume (V)	length × length × length	m^3	–
Density (ρ)	mass/volume	$kg\,m^{-3}$	–
Velocity (v)	displacement/time	$m\,s^{-1}$	–
Acceleration (a)	velocity/time	$m\,s^{-2}$	–
Force (F)	mass × acceleration	$kg\,m\,s^{-2}$	newton
Work (W)	force × distance	$kg\,m^2\,s^{-2}$	joule
Power (P)	work/time	$kg\,m^2\,s^{-3}$	watt
Charge (Q)	current × time	$A\,s$	coulomb
Voltage (V)	power/current	$kg\,m^2\,s^{-3}\,A^{-1}$	volt

Prefixes

In order to avoid writing too many zeroes when a quantity being measured is too small or too large, prefixes are used. Table 1.1.3 lists the commonly used prefixes.

Examples:

6.2 kilometres	6.2 km	$= 6.2 \times 10^3$ m
2.9 milliamperes	2.9 mA	$= 2.9 \times 10^{-3}$ A
4.1 micrometre	4.1 µm	$= 4.1 \times 10^{-6}$ m
100 picofarads	100 pF	$= 100 \times 10^{-12}$ F
3 megawatts	3 MW	$= 3 \times 10^6$ W

Table 1.1.3 *List of commonly used prefixes*

Prefix	Multiplying factor	Symbol
pico	10^{-12}	p
nano	10^{-9}	n
micro	10^{-6}	µ
milli	10^{-3}	m
centi	10^{-2}	c
deci	10^{-1}	d
kilo	10^3	k
mega	10^6	M
giga	10^9	G
tera	10^{12}	T

Conversion of units

Physics often requires that you convert from one unit to another. The key to converting between units is to first determine the relationship that exists between them. The following examples will illustrate this point.

Example

a Convert $80 \, \text{km h}^{-1}$ to m s^{-1}.

b Convert $2.5 \, \text{mm}^2$ to m^2.

c Convert $7.9 \, \text{g cm}^{-3}$ to kg m^{-3}.

a $80 \, \text{km h}^{-1} = \dfrac{80 \times 10^3 \, \text{m}}{1 \, \text{h}} = \dfrac{80 \times 10^3 \, \text{m}}{60 \times 60 \, \text{s}} = 22.2 \, \text{m s}^{-1}$

b $1 \, \text{mm}^2 = 1 \, \text{mm} \times 1 \, \text{mm} = 1 \times 10^{-3} \, \text{m} \times 1 \times 10^{-3} \, \text{m} = 1 \times 10^{-6} \, \text{m}^2$

$\therefore \quad 2.5 \, \text{mm}^2 = 2.5 \times 1 \times 10^{-6} \, \text{m}^2 = 2.5 \times 10^{-6} \, \text{m}^2$

c $7.9 \, \text{g cm}^{-3} = \dfrac{7.9 \, \text{g}}{1 \, \text{cm} \times 1 \, \text{cm} \times 1 \, \text{cm}} = \dfrac{7.9 \times 10^{-3} \, \text{kg}}{1 \times 10^{-6} \, \text{m}^3} = 7900 \, \text{kg m}^{-3}$

The mole and the Avogadro constant

One of the base quantities is 'the amount of substance'. The amount of substance can refer to the number of particles, number of molecules or number of ions. The number in this case refers to a value of 6.02×10^{23}. This number is called the **Avogadro constant** (N_A).

The **mole** is the amount of substance (n) that contains the same number of particles as there are in 12 grams of carbon-12.

$\therefore 1 \, \text{mol} = 6.02 \times 10^{23}$ particles

The amount of particles (N) present in an amount of substance (n) is given by $N = nN_A$.

Example

Calculate the number of molecules present in 1.2 moles of helium gas.

$1 \, \text{mol} = 6.02 \times 10^{23}$

$\therefore 1.2 \, \text{mol} = 1.2 \times 6.02 \times 10^{23} = 7.224 \times 10^{23}$ molecules of helium gas

Key points

- A physical quantity consists of the product of a numerical value and a unit.

- The base quantities are mass, length, time, temperature, current, amount of substance and luminous intensity.

- All other physical quantities are derived from the base quantities.

- Prefixes are used as shorthand, for writing very large or very small quantities.

- The mole is equivalent to 6.02×10^{23} particles.

1.2 Dimension and unit analysis

Learning outcomes

On completion of this section, you should be able to:

- recall the dimensions of base quantities
- determine the dimensions of derived quantities
- understand the importance of dimensional analysis.

Dimension of physical quantities

The dimension of a physical quantity shows the relation between the physical quantity and the base quantities listed in Table 1.1.1. Table 1.2.1 lists some of the dimensions of the base quantities.

Table 1.2.1 *Dimensions of base quantities*

Base physical quantity	Dimension
Mass	[M]
Length	[L]
Time	[T]
Temperature	[θ]
Electric current	[A]

The dimensions of derived physical quantities can be determined once the relationship between the corresponding base quantities is known. For example,

$$\text{density} = \frac{\text{mass}}{\text{volume}} = \frac{[M]}{[L^3]} = [ML^{-3}]$$

The dimension for density is therefore $[ML^{-3}]$.

Table 1.2.2 lists some derived physical quantities and their dimensions.

Table 1.2.2 *Dimensions of derived quantities*

Derived physical quantity	Dimension
Area	$[L]^2$
Volume	$[L]^3$
Density	$[ML^{-3}]$
Acceleration	$[LT^{-2}]$
Power	$[ML^2T^{-3}]$

Importance of dimensions and units

Dimensions can be used

1. To deduce the dimensions of a derived quantity (as shown previously).
2. To check the homogeneity of an equation.

In any scientific equation, the units on the left-hand side of the equation must equal to the units on the right-hand side of the equation. If the units are not the same, then the equation is incorrect.

Consider the equation $s = ut + \frac{1}{2}at^2$. It represents the displacement s of a body after time t, where u is initial velocity and a is acceleration (where a is constant). On the left-hand side of the equation the unit of displacement is the metre (m). On the right-hand side of the equation there are two terms.

The unit for ut is $m\,s^{-1} \times s = m$

The unit for $\frac{1}{2}at^2 = m\,s^{-2} \times s^2 = m$ (The coefficient $\frac{1}{2}$ is ignored.)

The unit on the right-hand side of the equation is m, even though the two terms are being added.

Since the unit on the left-hand side of the equation is equal to the units on the right-hand side of the equation, the equation is said to be homogeneous.

An equation that is not homogeneous is not correct.

An equation that is homogeneous may not necessarily be correct.

Suppose the same equation above is rewritten as $s = \frac{1}{2}at^2$. The units on both sides of the equation are the same (i.e. m). However, the equation is incorrect because it is missing a term.

3 To predict the form of equations.

Suppose that it is suggested that the velocity v of a wave on a stretched string is related to the tension in the string T, mass of the string m and length of the string l.

$v \propto T^x m^y l^z$

x, y and z are constants.

The units of $v = \mathrm{m\,s^{-1}}$ and the dimensions of $v = [\mathrm{L\,T^{-1}}]$

The units of $T = \mathrm{kg\,m\,s^{-2}}$ and the dimensions of $T = [\mathrm{M\,L\,T^{-2}}]$

The units of $m = \mathrm{kg}$ and the dimensions of $m = [\mathrm{M}]$

The units of $l = \mathrm{m}$ and the dimensions of $l = [\mathrm{L}]$

Considering the dimensions in the suggested relationship we get

$[\mathrm{L\,T^{-1}}] \propto [\mathrm{M\,L\,T^{-2}}]^x [\mathrm{M}]^y [\mathrm{L}]^z$

$[\mathrm{L\,T^{-1}}] \propto [\mathrm{M}]^{x+y} [\mathrm{L}]^{x+z} [\mathrm{T}]^{-2x}$

Now considering each dimension one at a time on the left-hand side and right-hand side, we get the following

M term $\quad 0 = x + y$

L term $\quad 1 = x + z$

T term $-1 = -2x$

Solving these equations we get $x = \frac{1}{2}$, $y = -\frac{1}{2}$, $z = \frac{1}{2}$.

The equation now becomes $v \propto T^{1/2} m^{-1/2} l^{1/2}$ or $v \propto \sqrt{\dfrac{Tl}{m}}$.

A limitation of this method of trying to predict the form of an equation is that it cannot determine the value of the constant of proportionality. This equation can only be verified experimentally.

Key points

- The dimension of mass, length and time are [M], [L] and [T] respectively.
- The dimension of derived quantities can be determined from base quantities.
- Dimensional analysis can be used to check the homogeneity of equations.
- An equation is incorrect if it is not homogenous.

✅ *Exam tip*

To check to see if an equation is correct, determine the units on the left-hand side and the right-hand side of the equation. If they are the same, the equation is homogeneous.

1.3 Scalar and vector quantities

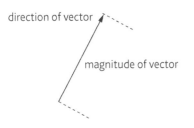

Figure 1.3.1 *representing a vector quantity*

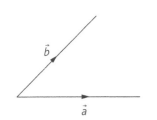

Figure 1.3.2

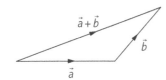

Figure 1.3.3 *Adding two vectors*

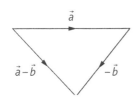

Figure 1.3.4 *Subtracting two vectors*

Scalar and vector quantities

Quantities can be classified as either being a scalar or a vector.

A **scalar quantity** has magnitude only.

A **vector quantity** has magnitude and direction.

Examples of scalar quantities are mass, length, work, speed, distance, **energy** and power.

Examples of vector quantities are weight, momentum, velocity, acceleration, and displacement (Figure 1.3.1).

A vector can be represented as a straight line with an arrow at one end. The length of the line represents the magnitude of the vector. The direction of the arrow points in the direction of the vector.

Adding and subtracting scalars

Scalar quantities are added and subtracted numerically. For example, if you were interested in finding the total mass of a student and his back pack you would perform the calculation as follows:

Mass of student = 50 kg, Mass of back pack = 4 kg

Total mass of student and his back pack = 50 + 4 = 54 kg

Suppose 90 J of energy is supplied to a machine. If the useful energy output of the machine is 60 J, determine the energy lost inside the machine.

Energy input = 90 J, Energy output = 60 J

Energy lost inside the machine = 90 – 60 = 30 J

Adding (combining) vectors

Adding vector quantities is not as simple as adding scalar quantities. Vector quantities have direction that has to be taken into account. Consider two vectors $\vec{a}$ and $\vec{b}$ as shown Figure 1.3.2.

In order to add these two vectors vector $\vec{a}$ is first drawn. Vector $\vec{b}$ is then drawn by starting from the point at which vector $\vec{a}$ ended. The vector $\vec{a} + \vec{b}$ is then drawn from the starting point of vector $\vec{a}$ to the ending point of vector $\vec{b}$. The vector $\vec{a} + \vec{b}$ is called the **resultant vector** (Figure 1.3.3).

Vector subtraction is a form of vector addition. For example the vector $\vec{a} - \vec{b}$ is the same as saying $\vec{a} + (-\vec{b})$. Using the previous example as guide to perform vector addition, vector $\vec{a}$ is first drawn. The vector $-\vec{b}$ is then drawn by starting from the point at which $\vec{a}$ ended. The vector $-\vec{b}$ is simply an arrow drawn having the same length as $\vec{b}$ but pointing in the opposite direction (Figure 1.3.4).

Vectors acting in the same direction

Force is a vector quantity. The unit of force is the newton (N). Suppose two forces are acting on an object in the same direction. The resultant force (combined effect of both forces) can be found by simply adding the magnitude of the two forces. In Figure 1.3.5 two forces 3 N and 5 N are acting in the same direction. The resultant force is 8 N to the right.

Vectors acting in opposite directions

Suppose two forces are acting on an object in opposite directions as shown in Figure 1.3.6. The resultant force is found by subtracting the magnitude of the two forces. The resultant force will act in the direction of the larger of the two forces. Therefore, the resultant force is 2 N to the right.

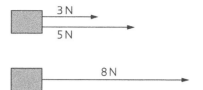

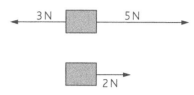

Figure 1.3.5 *Adding vectors acting in the same direction*

Vectors acting at an angle to each other

Suppose two forces are acting at an angle of θ to each other as shown in Figure 1.3.7. In order to add the two vectors, the force 4 N is drawn first. The 5 N force is then drawn by starting from the ending point of the 4 N force. The addition of the 4 N force and the 5 N force results in the force R. The value of the force R can be determined by scale drawing. A scale must first be decided upon. For example, 1 cm can represent 1 N. This means that the 4 N and 5 N force can be represented as lengths 4 cm and 5 cm respectively. Once the scale drawing is completed, the length R is measured using a ruler. Suppose $\theta = 60°$. The length of R will be 7.8 cm.

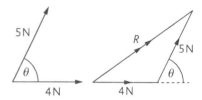

Figure 1.3.6 *Adding vectors acting in opposite directions*

Therefore, $R = 7.8\,\text{cm} \times 1\,\text{N}\,\text{cm}^{-1} = 7.8\,\text{N}$.

R can also be determined by calculation using the cosine rule.

$$c^2 = a^2 + b^2 - 2ab\cos\theta$$
$$c^2 = 5^2 + 4^2 - (2 \times 5 \times 4 \times \cos 120°) = 61$$
$$c = 7.8\,\text{N}$$

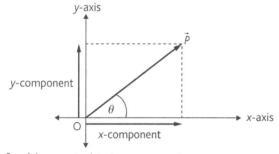

Figure 1.3.7 *Finding the resultant of two vectors acting at an angle to each other*

Resolving vectors

A vector can be replaced by two other vectors acting at right angles to each other. Consider the x–y plane shown in Figure 1.3.8. A vector $\vec{P}$ has been drawn using the origin as the starting point. The vector $\vec{P}$ can be replaced with two vectors acting at right angles to each other.

Figure 1.3.8 *Resolving a vector into two components*

The horizontal component (x-component) is found by drawing a vertical line from the tip of vector $\vec{P}$ parallel to the y-axis until it meets the x-axis.

The vertical component (y-component) is found by drawing a horizontal line from the tip of vector $\vec{P}$ parallel to the x-axis until it meets the y-axis.

Suppose a ball is struck such that it travels with a velocity of $5\,\text{m}\,\text{s}^{-1}$ at $30°$ to the horizontal as shown in Figure 1.3.9.

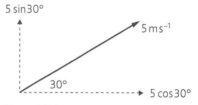

Figure 1.3.9

The horizontal and vertical components can be found by scale drawing or by calculation.

Horizontal component $= 5 \times \cos 30° = 4.33\,\text{m}\,\text{s}^{-1}$

Vertical component $= 5 \times \cos(90 - 30) = 5 \times \sin 30° = 2.5\,\text{m}\,\text{s}^{-1}$

Key points

- A scalar quantity has magnitude only.

- A vector quantity has a magnitude and a direction.

- Scalar quantities are added and subtracted algebraically.

- Vectors are added by taking into account their directions.

- Any vector can be resolved into two vectors which act at right angles to each other.

2.1 Measurements 1

Measuring lengths

The metre rule, vernier calliper and micrometre screw gauge are common instruments used to measure lengths in a laboratory. The SI unit of length is the metre (m). A metre rule would be used to measure the width of a desk or the length of a pendulum. A vernier calliper would be used to measure the dimensions of a small block of wood or the diameter of a test tube. A micrometre screw gauge would be used to measure the diameter of a piece of copper wire. The choice of measuring device depends on the magnitude of the length being measured.

Measuring length using a vernier calliper

Figure 2.1.1 shows a diagram of a vernier calliper. There is a main scale and a vernier scale. When an object is placed between outside jaws, the main scale is read first. On the vernier scale, one of the markings will line up with the main scale. This gives the fraction of the millimetre scale that must be added to the main scale. The vernier calliper in Figure 2.1.2 is read as follows:

$$\text{Reading} = \text{main scale} + \text{vernier scale} = 56 + 0.7 = 56.7\,\text{mm}$$

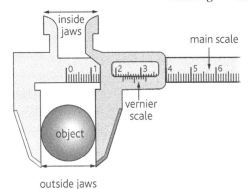

Figure 2.1.1 *A vernier calliper*

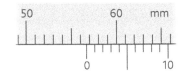

Figure 2.1.2 *Reading a vernier calliper*

Measuring length using a micrometre screw gauge

Figure 2.1.3 shows a diagram of a micrometre screw gauge. It consists of a main scale on the shaft and a fractional scale on a rotating barrel. The fractional scale has 50 divisions. One complete turn represents 0.50 mm. The micrometre screw gauge in Figure 2.1.4 is read as follows:

$$\text{Reading} = \text{main scale} + \text{rotating scale} = 6.50 + 0.23 = 6.73\,\text{mm}$$

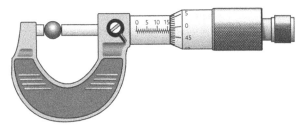

Figure 2.1.3 *A micrometre screw gauge*

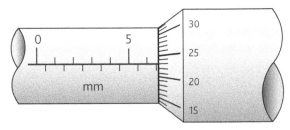

Figure 2.1.4 *Reading a micrometre screw gauge*

Measuring mass and weight

The mass of an object is measured using a beam balance (Figure 2.1.5).

The weight of an object is the force exerted on it by gravity. Weight can therefore be measured using a spring balance (Figure 2.1.6).

Measuring angles

It often required that angles be measured in some experiments. Angles can be measured by taking accurate measurements of lengths and using trigonometric calculations to determine angles. Where possible, angles can be measured directly using protractors. Measurement of angles is often required when performing ray optic experiments or demonstrating the equilibrium of forces.

There are optics experiments that require very precise measurements of angles. In these experiments a spectrometer is used. Figure 2.1.7 illustrates a spectrometer.

Measuring temperature

Temperature is measured using a thermometer. The SI unit of temperature is the kelvin (K). Temperature is also measured in degrees Celsius (°C). (Refer to 14.2, *Thermometers*.)

Measuring volume

The volume of an object is the amount of space taken up by the object. The volume of regular objects can be determined by calculation.

Volume of a cuboid $V = l \times b \times h$ (length l, breadth b, height h)

Volume of a sphere $V = \frac{4}{3}\pi r^3$ (radius of sphere r)

Volume of cylinder $V = \pi r^2 h$
(radius of base of cylinder r, height of cylinder h)

Volume is commonly measured in cm^3 or m^3.

The volume of an irregularly shaped object can be measured using a displacement method. Suppose you are required to measure the volume of a small stone. Some water is place into a measuring cylinder and the initial volume recorded. The stone is gently placed into the water and the final volume recorded. The difference between the two volumes gives the volume of the stone (Figure 2.1.8).

Key points

- Standard instruments used to measure lengths are the metre rule, vernier calliper and the micrometre screw gauge.
- Mass is measured using a beam balance or an electronic balance.
- Weight is measured using a spring balance.
- Angles are measured using a protractor.
- A spectrometer is used when measuring angles in optical experiments.
- Temperatures are measured using a thermometer.
- The volume of a regular object is determined by calculation.
- The volume of an irregularly shaped object is determined using a displacement method.

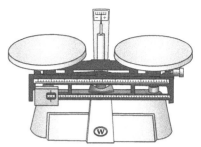

Figure 2.1.5 *A beam balance*

Figure 2.1.6 *A spring balance*

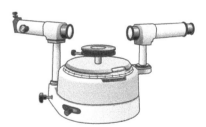

Figure 2.1.7 *A spectrometer*

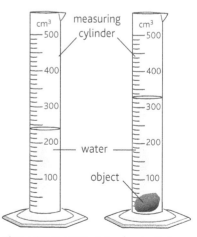

Figure 2.1.8 *Measuring volume using a displacement method*

Learning outcomes

On completion of this section, you should be able to:

- measure time using a clock, stopwatch and the time base of a cathode-ray oscilloscope

- measure electrical quantities using standard laboratory instruments

- understand how to use calibration curves

- understand how to rearrange an equation into a linear form.

Measuring time

The SI unit of time is the second (s). Time durations are measured using clocks or stopwatches. Suppose an experiment is performed to measure the time taken for a pendulum to complete one oscillation. In order to get an accurate value for this time interval, the time taken for 10 oscillations T_{10} is recorded using a stopwatch. The experiment is repeated several times and the mean (average) time for ten oscillations is recorded. The time for one oscillation T is determined as follows: $T = T_{10}/10$.

There are instances when the time interval of an event is so small that a stopwatch cannot be used. An instrument called a cathode-ray oscilloscope (CRO) can be used. The CRO consists of a pair of parallel metal plates inside it called the X-plates. A sweep generator of known frequency is attached to the X-plates. This frequency is adjusted using the time-base setting on the front panel of the CRO (Figure 2.2.1).

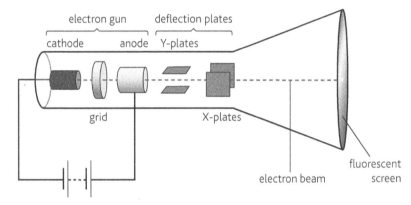

Figure 2.2.1 *A simple diagram of a cathode-ray oscilloscope*

Figure 2.2.2 shows a waveform on the screen of a CRO. The time-base setting is calibrated at $2\,\mathrm{ms\,cm^{-1}}$. Each square on the screen is $1\,\mathrm{cm} \times 1\,\mathrm{cm}$. Suppose it is required that the time interval between A and B be found. The distance AB is $6\,\mathrm{cm}$. Therefore, the time interval between A and B is $6\,\mathrm{cm} \times 2\,\mathrm{ms\,cm^{-1}} = 12\,\mathrm{ms}$.

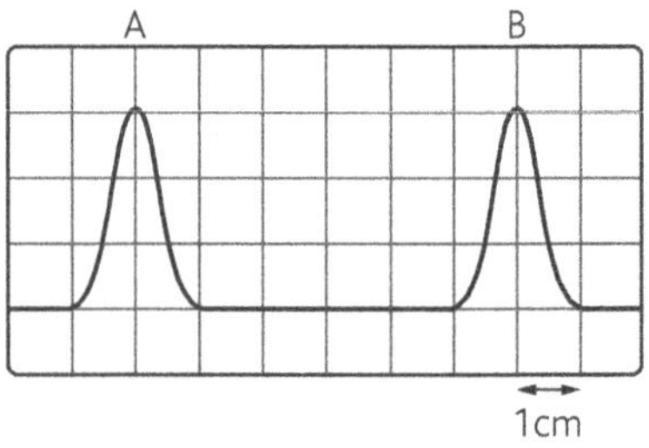

Figure 2.2.2 *A waveform on the screen of a CRO*

Measuring electrical quantities

Two important electrical quantities are **electric current** and **potential difference**. An electric current is measured using an instrument called an ammeter and its unit is the ampere (A). A potential difference is measured using an instrument called a voltmeter and its unit is the volt (V). The ammeter and the voltmeter can be an analogue or digital type. In the case of the analogue-type meter the location of the pointer in reference to a scale is recorded. In the case of a digital meter, the measurement is recorded as seen on the display of the instrument.

Calibration curves

Suppose you were provided with a mercury-in-glass thermometer (A) that had no markings on the length of it. The thermometer is of no use if it is not calibrated. The thermometer is placed in known temperatures

(temperature of pure melting ice 0 °C and the temperature of steam 100 °C above pure boiling water) and the length of mercury is measured. These temperatures are easily reproducible and chosen for this reason. Another thermometer (B) which is already calibrated is used for comparison. Both thermometers are placed in substances that have temperatures between 0 °C and 100 °C. The temperature reading on the calibrated thermometer (B) is recorded and the length of the mercury in the thermometer (A) is recorded.

A graph of temperature against length of mercury is then plotted (Figure 2.2.3). This graph is the calibration curve for the thermometer A. When the thermometer A is placed in a substance of unknown temperature, the length of mercury is recorded. This length is read off from the calibration curve to determine the unknown temperature.

Plotting linear graphs from non-linear relationships

In practical work it is often required to establish relationships between two quantities. If two quantities x and y are related such that they have a linear relationship, a straight line graph would be obtained when y is plotted against x.

The equation of a straight line is of the form $y = mx + c$, where m is the gradient of the straight line and c is the intercept on the y-axis.

It is often required that an expression be re-written so that it resembles the form of the equation of a straight line. Table 2.2.1 shows some examples.

Table 2.2.1

Expression	What to plot?	constants	Gradient	y-intercept
$y = ax^2 + b$	y against x^2	a and b	a	$(0, b)$
$T = kl^n$	$\lg T$ against $\lg l$	k and l	n	$(0, \lg k)$
$y^2 = ax^2 + bx$	$\dfrac{y^2}{x}$ against x	a and b	a	$(0, b)$
$N = Ae^{-kt}$	$\ln N$ against t	A and k	$-k$	$(0, \ln A)$

Suppose T and l are related by the following equation $T = kl^n$.

$\lg T = \lg (kl^n)$ Take $\log_{10}$ on both sides of the equation

$\lg T = \lg k + \lg (l^n)$

$\lg T = \lg k + n \lg l$ ⟵ Linear form

A plot of a graph of $\lg T$ against $\lg l$ will produce a straight line. The gradient is n and the y-intercept is $\lg k$.

Suppose N and t are related by the following equation $N = Ae^{-kt}$.

$\ln N = \ln (Ae^{-kt})$ Take $\log_e$ on both sides of the equation

$\ln N = \ln A + \ln (e^{-kt})$

$\ln N = \ln A - kt \ln e$

$\ln N = \ln A - kt$ ⟵ Linear form

A plot of a graph of $\ln N$ against t will produce a straight line. The gradient is $-k$ and the y-intercept is $\ln A$.

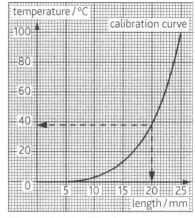

Figure 2.2.3 A calibration curve

☑ *Exam tip*

Recall the rules for logarithms

$$\log_b (A)^n = n \log_b A$$
$$\log_b (A) + \log_b (B) = \log_b (AB)$$
$$\log_b (A) - \log_b (B) = \log_b \left(\frac{A}{B}\right)$$

☑ *Exam tip*

$\log_{10}$ is usually written as $\lg$.

$\log_e$ is usually written as $\ln$, where $e = 2.718$

Key points

- Time is measured using a clock or stopwatch.
- A cathode-ray oscilloscope can be used to measure very small time intervals.
- Electrical current is measured using an ammeter.
- Electrical voltage is measured using a voltmeter.
- Calibration curves are used to calibrate instruments.
- Expressions involving two quantities can be rearranged in such a way that a linear graph can be plotted.

2.3 Errors in measurements

Learning outcomes

On completion of this section, you should be able to:

- differentiate between systematic and random errors
- identify ways of reducing systematic and random errors
- differentiate between precision and accuracy.

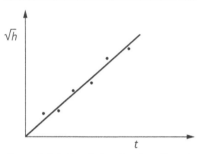

Figure 2.3.1 *Graph showing the effect of random errors*

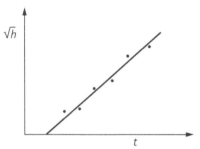

Figure 2.3.2 *Graph showing the effect of a systematic error*

Systematic and random errors

Whenever a physical quantity is measured, there is the likelihood that there will be an error or uncertainty in the measurement. Errors can be divided into **systematic** and **random errors**. If a physical quantity is measured a large number of times, it may be revealed that the readings fluctuate around some value. Some readings may be larger or smaller. This type of error is known as a random error and usually occurs as a result of the experimenter. Suppose a student measures the time taken t, for a steel ball to fall from rest through a known vertical distance h. The distance h is varied and the corresponding time is measured.

It is known that h and t are related by the following equation

$h = \frac{1}{2}gt^2$, where g is the acceleration due to free fall $(g = 9.81\,\mathrm{m\,s^{-2}})$.

According to this relationship, a plot of $\sqrt{h}$ against t will give a straight line through the origin. The graph in Figure 2.3.1 shows a plot of $\sqrt{h}$ against t.

All the data points do not lie on the straight line. They are scattered above and below the line. The deviation of the points from the line indicates that there are random errors in the experiment. Random errors can be reduced by repeating measurements and finding the mean of the measurements. In this experiment, the student could have taken several measurements of t for a given height h, and find the average of those times.

Suppose a different student performs the same experiment. The results obtained are shown in Figure 2.3.2.

All the data points are scattered about the main line. This illustrates a random error. However, notice that the graph does not pass through the origin. This indicates that there is a systematic error in the experiment. In the case of a systematic error, there is a constant error in one direction. Either all the readings are larger than their true value or all the readings are smaller than their true value. A systematic error can occur because of

- A zero error in the instrument being used. In this case, the instrument gives a reading when the physical quantity is not present. This error can be eliminated by zeroing the instrument, if possible, before performing the experiment.
- An incorrectly calibrated instrument. In this case, the instrument may have been used for a long period of time and the readings are no longer accurate. Instruments are to be calibrated as often as recommended by the manufacturer of the instrument.
- Improper techniques being used by the experimenter to perform the experiment. If the experimenter consistently makes the same mistake when measuring a quantity, all the results will be off by the same amount. For example, suppose an experimenter accidentally assumes that the smallest division on a scale is 0.2 m when it is actually 0.1 m. A reading of 1.7 m may be recorded as 1.6 m.

Precision and accuracy

An easy way to understand the difference between **precision** and **accuracy** is to consider a game of darts. The objective of the game is to hit the bull's eye with the darts. Figure 2.3.3 illustrates the various scenarios.

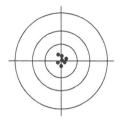 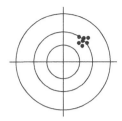 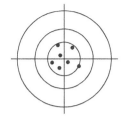 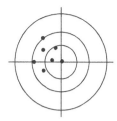

a *Precise and accurate*   **b** *Precise but not accurate*   **c** *Accurate but not precise*   **d** *Not precise and not accurate*

Figure 2.3.3 *A game of darts*

Notice that accuracy has to do with how close the darts are to the bull's eye.

Precision has to do with how close the darts are as a group.

Suppose a quantity has a true value of x_0. A student measures the quantity a large number of times n using an appropriate instrument. The experiment is known to have systematic and random errors. A graph is plotted to show the number of times n, a particular reading x, is obtained. Figure 2.3.4 illustrates various scenarios. The graphs help illustrate the difference between precision and accuracy.

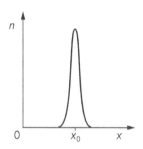 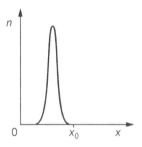 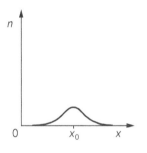 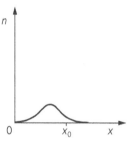

a *Precise and accurate*   **a** *Precise but not accurate*   **a** *Accurate but not precise*   **a** *Not precise and not accurate*

Figure 2.3.4 *Experiments having systematic and random errors*

- Measurements are accurate if the systematic errors are small.
- Measurements are precise if the random errors are small.

Example

Suppose two students perform an experiment to determine the value for the acceleration due to gravity g. g is known to be $9.81\,\mathrm{m\,s^{-2}}$. The students repeat the experiment several times and the value of g is recorded.

    Student A – 9.80, 9.82, 9.83, 9.81, 9.82

    Student B – 8.45, 8.41, 8.42, 8.45, 8.43

The results of the experiments performed by student A are both accurate and precise.

The results of the experiments performed by student B are precise but not accurate.

Key points

- Random errors occur as a result of the experimenter and can result in an error that is above or below the true value.

- Random errors can be reduced by repeating measurements and finding the mean.

- Random errors cannot totally be eliminated.

- Systematic errors are constant errors in one direction.

- Systematic errors can be eliminated.

- Precision is a measure of the reproducibility of a result.

- Accuracy is a measure of the closeness of the measured value to the true value.

Learning outcomes

On completion of this section, you should be able to:

- understand the terms *absolute error*, *fractional error* and *percentage error*
- calculate the uncertainties in derived quantities.

Uncertainties in derived quantities

When a quantity is measured, there is an error or uncertainty associated with the measurement. Suppose a metre rule is used to measure the length of a metal rod. The metre rule is able to give readings to nearest 0.1 cm. This means that when a reading is taken, the experimenter will either record to the nearest marking above or below the actual length of the rod. The reading will be either 0.05 cm too high or too low from the actual value. This error in the measurement is written as ±0.05 cm (half the smallest reading on the metre rule). If a student measures the length of the rod as 12.40 cm, then the information is recorded as follows:

12.40 ± 0.05 cm.

The actual or absolute uncertainty (error) is 0.05 cm.

The fractional or relative uncertainty (error) is $\dfrac{0.05}{12.40} = 0.004$.

The percentage uncertainty is $\dfrac{0.05}{12.40} \times 100\% = 0.4\%$.

Adding and subtracting

Consider two quantities P and Q.

$$P = 25.10 \pm 0.05$$

$$Q = 62.50 \pm 0.05$$

The absolute uncertainty in P is 0.05

The fractional uncertainty in P is $\dfrac{0.05}{25.10} = 0.002$

The percentage uncertainty in P is $\dfrac{0.05}{25.10} \times 100\% = 0.2\%$

The absolute uncertainty in Q is 0.05

The fractional uncertainty in Q is $\dfrac{0.05}{62.50} = 0.0008$

The percentage uncertainty in Q is $\dfrac{0.05}{62.50} \times 100\% = 0.08\%$

Suppose it is required to find the absolute uncertainty in the quantity $P + Q$ and $Q - P$.

$$P + Q = 25.1 + 62.5 = 87.6$$

The absolute uncertainty is found by adding the absolute uncertainties of each quantity.

Therefore, the absolute uncertainty in $P + Q$ is 0.05 + 0.05 = 0.1.

$$\therefore \qquad P + Q = 87.6 \pm 0.1$$

$$Q - P = 62.5 - 25.1 = 37.4$$

The absolute uncertainty is found by adding the absolute uncertainties of each quantity.

Therefore, the absolute uncertainty in $Q - P$ is 0.05 + 0.05 = 0.1.

$$\therefore \qquad Q - P = 37.4 \pm 0.1$$

Multiplication, division, powers and roots

Whenever quantities are multiplied or divided the percentage uncertainty is found by adding the percentage uncertainties of each of the quantities involved.

Whenever a quantity is raised to a power, the percentage uncertainty is found by multiplying the power by the percentage uncertainty of the quantity involved.

Whenever the nth root of a quantity is being found, the percentage uncertainty is found by multiplying $1/n$ by the percentage uncertainty of the quantity involved.

Consider the following quantities:

$$P = 18.2 \pm 0.1$$

$$Q = 6.24 \pm 0.02$$

Determine the percentage uncertainties in:

1 $P \times Q$ **2** $\dfrac{P}{Q}$

3 P^2 **4** $\sqrt[3]{Q}$

Percentage uncertainty in P is $\dfrac{0.1}{18.2} \times 100\% = 0.5\%$

Percentage uncertainty in Q is $\dfrac{0.02}{6.24} \times 100\% = 0.3\%$

1 The percentage uncertainty in $P \times Q$ is $0.5 + 0.3 = 0.8\%$

2 The percentage uncertainty in $\dfrac{P}{Q}$ is $0.5 + 0.3 = 0.8\%$

3 The percentage uncertainty in P^2 is $2 \times 0.5 = 1\%$

4 The percentage uncertainty in $\sqrt[3]{Q}$ is $\dfrac{1}{3} \times 0.3 = 0.1\%$

Example

1 A student wishes to measure the volume V of a wire of length l and obtains the following measurements.

Diameter of the wire $d = (0.94 \pm 0.04)\,\text{mm}$

Length $l = (839 \pm 3)\,\text{mm}$

The volume V of the wire is given by $\frac{1}{4}\pi d^2 l$

Calculate the percentage uncertainty in the measurement of:

i the diameter d

ii the length l

iii the volume V.

i % uncertainty in the measurement of $d = \dfrac{\Delta d}{d} = \dfrac{0.04}{0.94} \times 100\%$
$= 4\%$

ii % uncertainty in the measurement of $l = \dfrac{\Delta l}{l} = \dfrac{3}{839} \times 100\%$
$= 0.4\%$

iii % uncertainty in the measurement of $V = 2\dfrac{\Delta d}{d} + \dfrac{\Delta l}{l} = 2(4) + 0.4$
$= 8.4\%$

Exam tip

Ensure that you understand the worked examples. Uncertainty questions pose a challenge to students.

Key points

- Measured quantities have an error associated with them known as the absolute error.

- The absolute error is usually taken as half the smallest reading on the instrument scale.

- The fractional error is the absolute error divided by the measured value.

- When quantities are added or subtracted, the absolute uncertainties are added.

- When quantities are multiplied or divided, the fractional uncertainties of the quantities are added.

- When a quantity is raised to a power, the fractional uncertainty of the quantity is multiplied by the power.

3 Kinematics

3.1 Kinematics

Kinematics is the term used to describe the motion of objects without considering what is actually causing the motion.

Distance and displacement

Suppose a dog is playing in a park. He starts at the point A, moves around the park and ends up at the point B. The dotted line in Figure 3.1.1 shows the path taken by the dog. This path gives the distance travelled by the dog. The arrow joining A and B represents the **displacement** of the dog. Distance is a scalar quantity and has a magnitude only. Displacement is, however, a vector quantity. This means that it has a magnitude as well as a direction. It is possible for the dog to have zero displacement after his walk around the park. If the dog had returned to the point A (starting point) his displacement would be zero, but his distance travelled would have a non-zero magnitude.

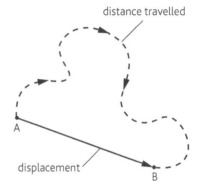

Figure 3.1.1 *Differentiating between distance and displacement*

Speed and velocity

Speed is defined as the rate of change of distance. Speed is a scalar quantity. If an object changes its speed several times during its journey, you might be interested in its average speed. The average speed is the total distance travelled divided by the total time taken to complete the journey. The SI unit of speed is metres per second (m/s or m s^{-1}).

Definition

Speed is the rate of change of distance.

Definition

Average speed is equal to half of the initial speed plus the final speed.

Equation

$$v = \frac{s}{t}$$

v – speed/m s^{-1}
s – distance travelled/m
t – time/s

Equation

$$\text{Average speed} = \frac{u + v}{2}$$

u – initial speed/m s^{-1}
v – final speed/m s^{-1}

Velocity is defined as the rate of change of displacement. Velocity is a vector quantity. The SI unit of velocity is metres per second ($m\,s^{-1}$).

Definition	Equation
Velocity is the rate of change of displacement.	$v = \dfrac{\Delta s}{\Delta t}$ v – velocity/$m\,s^{-1}$ Δs – change in displacement/m Δt – change in time/s

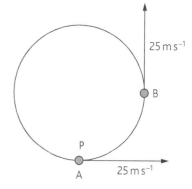

Figure 3.1.2 *An object moving in a circular path*

Consider an object P, moving in a circular path at a constant speed of $25\,m\,s^{-1}$ (Figure 3.1.2). At any point on the circle, the speed of the object is $25\,m\,s^{-1}$. The velocity at the points A and B are different since the direction of motion has changed.

Acceleration

Acceleration is defined as the rate of change of velocity. An object accelerates when its velocity changes. Acceleration is a vector quantity and therefore the direction of motion of the object is taken into consideration. The SI unit of acceleration is metres per second squared ($m\,s^{-2}$).

Definition	Equation
Acceleration is the rate of change of velocity.	$a = \dfrac{v - u}{t}$ a – acceleration/$m\,s^{-2}$ u – initial velocity/$m\,s^{-1}$ v – final velocity/$m\,s^{-1}$ t – time/s

☑ *Exam tip*

When the velocity of an object increases, the acceleration is positive. When the velocity of an object decreases, the acceleration is negative and is referred to as a deceleration.

Graphical representation of motion in a straight line

The motion of an object moving in a straight line can be represented by a graph. These graphs are particularly useful when the object is moving with non-uniform acceleration. If the acceleration is uniform, the equations of motion can be used to analyse motion. If the acceleration is not uniform, graphical methods are used to analyse motion.

Displacement–time graphs

Suppose an object P is stationary at a point O. It then moves with a constant or uniform velocity in a straight line for some time. The displacement–time graph in Figure 3.1.3 shows the motion of the object. The graph is a straight line through the origin. The velocity is determined by finding the gradient of the straight line.

Suppose an object is dropped from a height of 2 m above the ground. Figure 3.1.4 shows the displacement–time graph for the motion. The graph for this motion is a curve. The object is accelerating uniformly at a rate of $9.81\,m\,s^{-2}$. The velocity at any point on the curve is determined by finding the gradient of a tangent drawn at that point on the curve. The velocity at the point P on the curve is determined by first drawing a tangent at that point. A tangent is a straight line drawn such that it just touches the curve at the point P. The gradient of the tangent is equal to the velocity at the point P.

Figure 3.1.3 *Constant velocity*

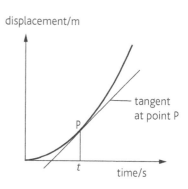

Figure 3.1.4 *Non-uniform velocity*

Velocity–time graphs

The motion of an object can be represented using a velocity–time graph. The area under a velocity–time graph measures the displacement of the object. The gradient of a velocity–time graph measures the acceleration of the object.

Consider Figure 3.1.5. It represents a velocity–time graph.

The various sections of the graph can be described as follows.

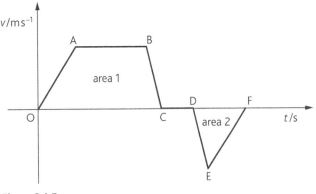

OA – The object accelerates uniformly from rest.

AB – The object is travelling at constant velocity.

BC – The object decelerates uniformly.

CD – The object is stationary. Its velocity is zero.

DE – The object reverses direction and accelerates uniformly from rest.

EF – The object decelerates uniformly and then comes to rest.

Figure 3.1.5

The acceleration at any point in the journey is found by finding the gradient of the curve at that point.

The area under a velocity–time graph measures displacement.

Area 1 gives the displacement of the object from O to C.

Area 2 gives the displacement of the object from D to F.

Example

Figure 3.1.6 shows the velocity–time graph for a journey lasting 50 seconds.

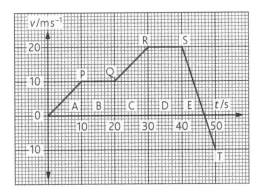

Figure 3.1.6

Use the information from the graph to find:

i   the velocity 10 seconds after the start of the journey

ii the acceleration during the first 10 seconds

iii the acceleration between 40 and 45 seconds

iv the distance travelled between 10 and 20 seconds

v   the distance travelled between 20 and 40 seconds.

i $v = 10\,\mathrm{m\,s^{-1}}$

ii acceleration $=$ gradient of line OP $= \dfrac{10 - 0}{10 - 0} = 1\,\mathrm{m\,s^{-2}}$

iii acceleration $=$ gradient of line ST $= \dfrac{20 - (-10)}{40 - 50} = -3\,\mathrm{m\,s^{-2}}$

iv Distance travelled $=$ Area B $= 10 \times 10 = 100\,m$

v Distance travelled $=$ Area C $+$ Area D

$\quad = \dfrac{1}{2}(10 + 20)10 + (20 \times 10) = 150 + 200 = 350\,\mathrm{m}$

Example

Suppose a ball is dropped from a fixed height above the ground on a metal surface and bounces several times. Sketch the velocity–time graph and hence deduce the acceleration–time graph for the motion of the ball.

The initial velocity of the ball is zero. As the ball falls, its velocity increases uniformly because gravity makes it accelerate at a constant rate of $9.81\,\mathrm{m\,s^{-1}}$. When it hits the metal surface, it changes direction in a short time interval. Its velocity decreases until it is zero at the highest point. The ball changes direction again and its velocity increases again. The velocity–time graph for the motion of the ball is shown in Figure 3.1.7. Acceleration is defined as the rate of change of velocity. The gradient of a velocity–time graph gives acceleration. The straight line portions of the velocity–time graph represent constant acceleration.

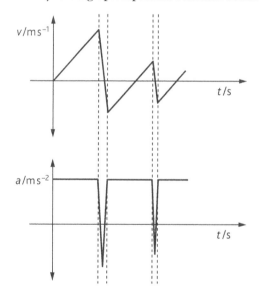

Figure 3.1.7

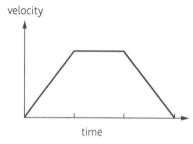

Figure 3.1.8

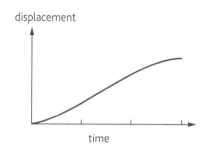

Figure 3.1.9

Example

Figure 3.1.8 shows velocity–time graph for the motion of a car. Describe the motion of the car and sketch the displacement–time graph for the motion of the car.

The car accelerates uniformly from rest. The car then travels a constant velocity for some time. The car then decelerates uniformly until it comes to rest.

Figure 3.1.9 shows the displacement–time graph.

Key points

- Displacement is the distance moved in a particular direction.
- Speed is a scalar quantity.
- Velocity and acceleration are vector quantities.
- Velocity is the rate of change of displacement.
- Acceleration is the rate of change of velocity.
- The slope of a displacement–time graph measures velocity.
- The slope of a velocity–time graph measures acceleration.
- The area under a velocity–time graph measures displacement.

3.2 Equations of motion

Learning outcomes

On completion of this section, you should be able to:

- derive the equations of motion
- use the equations of motion to solve problems.

Derivation of the equations of motion

Consider an object P, initially travelling at u m s^{-1}. It then accelerates uniformly at a rate of a m s^{-2} to achieve a final velocity of v m s^{-1}. It takes t seconds to do so and P travels through a distance of s metres (Figure 3.2.1).

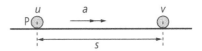

Figure 3.2.1 *An object P moving with uniform acceleration*

When deriving the equations of motion, the following assumptions are made.

1 The acceleration is uniform (constant).

2 The motion is in a straight line.

The velocity–time graph for the motion of P is shown in Figure 3.2.2.

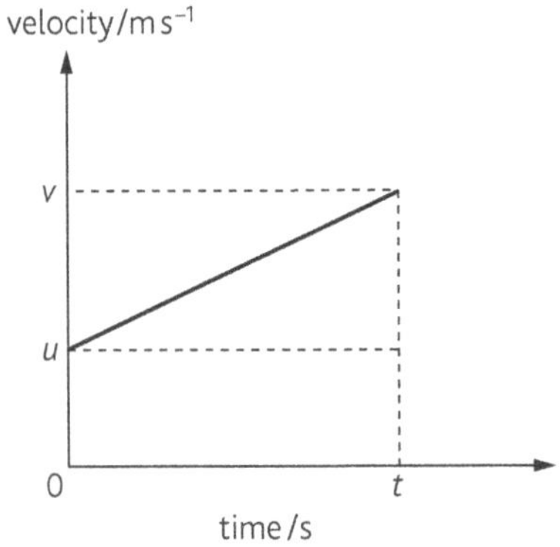

Acceleration is defined as the rate of change of velocity.

Therefore, $a = \dfrac{v - u}{t}$

Rearranging this equation we get the following:

$$\therefore \qquad v = u + at \qquad\qquad (1)$$

The displacement of P is given by s and is the area under the v–t graph.

$$\text{Displacement} = \text{area under graph}$$
$$= \text{area of rectangle} + \text{area of triangle}$$
$$s = (u \times t) + \left| \frac{1}{2} \times t(v - u) \right|$$

But from equation 1, $v - u = at$

$$s = (u \times t) + \left| \frac{1}{2} \times t(at) \right|$$
$$\therefore \qquad s = ut + \frac{1}{2}at^2 \qquad\qquad (2)$$

From Equation 1, $t = \dfrac{v - u}{a}$

Substituting this into Equation (2) we get the following:

$$\therefore s = u\left| \frac{v - u}{a} \right| + \frac{1}{2}a\left| \frac{v - u}{a} \right|^2$$

$$\therefore s = \frac{2u(v - u) + (v - u)^2}{2a} = \frac{2uv - 2u^2 + v^2 - 2uv + u^2}{2a} = \frac{v^2 - u^2}{2a}$$

Rearranging, we get the following:

$$v^2 - u^2 = 2as$$
$$v^2 = u^2 + 2as \qquad\qquad (3)$$

velocity/m s^{-1}

Figure 3.2.2 *Velocity–time graph for the motion of P*

☑ Exam tip

Ensure that you know the conditions under which the equations of motions are applicable.

Using the equations of motion

Example

A car starts from rest and accelerates at a rate of $2.5\,\mathrm{m\,s^{-2}}$ for $5.2\,\mathrm{s}$. It maintains a constant speed for $90\,\mathrm{s}$. The brake is then applied and the car comes to rest in $6\,\mathrm{s}$.

Calculate:

i the maximum velocity of the car

ii the total distance travelled.

---

i $v = u + at = 0 + (2.5 \times 5.2) = 13\,\mathrm{m\,s^{-1}}$

ii Distance travelled during the acceleration stage is determined as follows:

$$u = 0\,\mathrm{m\,s^{-1}}, a = 2.5\,\mathrm{m\,s^{-2}}, t = 5.2\,\mathrm{s}$$

$$s = ut + \frac{1}{2}at^2 = (0 \times 5.2) + \frac{1}{2}(2.5)(5.2)^2 = 33.8\,\mathrm{m}$$

Distance travelled during the stage where the velocity is constant:

$$u = 13\,\mathrm{m\,s^{-1}}, t = 90\,\mathrm{s}, a = 0\,\mathrm{m\,s^{-2}}$$

$$s = ut + \frac{1}{2}at^2 = (13 \times 90) + \frac{1}{2}(0)(90)^2 = 1170\,\mathrm{m}$$

Distance travelled during the deceleration stage:

$$u = 13\,\mathrm{m\,s^{-1}}, t = 6\,\mathrm{s}, v = 0\,\mathrm{m\,s^{-1}}$$

$$a = \frac{v - u}{t} = \frac{0 - 13}{6} = -2.17\,\mathrm{m\,s^{-2}}$$

$$v^2 = u^2 + 2as$$

$$(0)^2 = (13)^2 + 2(-2.17)\,s$$

$$s = 38.9\,\mathrm{m}$$

Total distance travelled $= 33.8 + 1170 + 38.9 = 1242.7\,\mathrm{m}$

Example

A ball is thrown vertically upwards with an initial velocity of $12\,\mathrm{m\,s^{-1}}$. Neglecting air resistance, determine:

i the distance travelled by the ball after 0.8 seconds

ii the velocity of the ball after 0.8 seconds

iii the time taken for the ball to reach its maximum height

iv the maximum height reached by the ball.

[$g = 9.81\,\mathrm{m\,s^{-2}}$]

Since the ball is thrown upwards, it decelerates. The acceleration of the ball is $-9.81\,\mathrm{m\,s^{-2}}$.

i $s = ut + \frac{1}{2}at^2 = (12 \times 0.8) + \frac{1}{2}(-9.81)(0.8)^2 = 6.46\,\mathrm{m}$

ii $v = u + at = 12 + (-9.81 \times 0.8) = 4.15\,\mathrm{m\,s^{-1}}$

iii At the maximum height $v = 0$

$$v = u + at$$

$$0 = 12 + (-9.81 \times t)$$

$$t = \frac{12}{9.81} = 1.22\,\mathrm{s}$$

iv $v^2 = u^2 + 2as$

$$(0)^2 = (12)^2 + 2(-9.81)\,s$$

$$s = \frac{(12)^2}{2 \times 9.81} = 7.33\,\mathrm{m}$$

✔ Exam tip

When an object is thrown vertically upwards and air resistance is ignored, the acceleration is equal to $-g$ $(-9.81\,\mathrm{m\,s^{-2}})$.

When an object is released and falls vertically downwards, its acceleration is g $(9.81\,\mathrm{m\,s^{-2}})$ if air resistance is ignored.

Key point

- The equations of motion are used for objects travelling in a straight line at a constant acceleration.

3.3 Projectile motion

Learning outcomes

On completion of this section, you should be able to:

- show that the path taken by a projectile is parabolic
- perform calculations involving projectile motion.

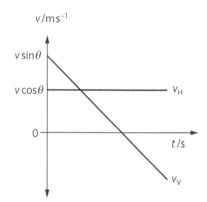

Figure 3.3.1 *Variation of the horizontal and vertical components with time*

Projectile motion

Suppose a ball is projected with a velocity V at an angle of θ to the horizontal.

The horizontal component is given by $V_H = V\cos\theta$.

The vertical component is given by $V_V = V\sin\theta$.

Figure 3.3.1 shows the variation with time of the horizontal and vertical components. The force of gravity acts vertically and only affects the vertical component of the ball. The horizontal component is unaffected by gravity and remains constant, provided that air resistance is ignored.

If air resistance is taken into account, the horizontal component is no longer constant.

The path taken by the ball is parabolic as shown in Figure 3.3.2.

In order to analyse projectile motion, the horizontal and vertical motion are treated separately.

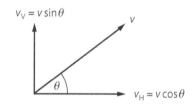

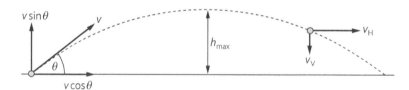

Figure 3.3.2 *A projectile*

Showing that the path taken by a projectile is parabolic

Object projected horizontally (Figure 3.3.3)

Suppose a ball is projected horizontally with a velocity v at a height h above the ground.

The horizontal component $v_H = v$

If air resistance is ignored, this component remains constant. Therefore, the acceleration is zero.

The horizontal displacement x at time t is given by

$$x = (v)t + \frac{1}{2}(0)t^2 = (v)t$$

$$\therefore \quad t = \frac{x}{v} \tag{1}$$

Initial vertical component of velocity $v_V = 0$

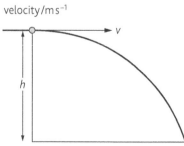

Figure 3.3.3 *An object projected horizontally*

The vertical displacement y at time t is given by

$$y = (0)t + \frac{1}{2}(-g)t^2 \tag{2}$$

[Assuming that positive velocity means that the ball is moving upwards, the acceleration due to gravity becomes $-g$.]

Substituting Equation (1) into Equation (2):

$$y = (0)\left|\frac{x}{v}\right| + \frac{1}{2}(-g)\left|\frac{x}{v}\right|^2$$

$$\therefore \quad y = -\left(\frac{g}{2v^2}\right)x^2$$

This equation is a parabola of the form $y = -ax^2$

Object projected at an angle (Figure 3.3.2)

Suppose a ball is projected with a velocity v at an angle of θ to the horizontal.

The horizontal component of velocity $v_H = v\cos\theta$

If air resistance is ignored, this component remains constant. Therefore, the acceleration is zero.

The horizontal displacement x at time t is given by

$$x = (v\cos\theta)t + \frac{1}{2}(0)t^2 = (v\cos\theta)t$$

$$\therefore \quad t = \frac{x}{v\cos\theta} \tag{3}$$

Initial vertical component of velocity $v_V = v\sin\theta$

The vertical displacement y at time t is given by

$$y = (v\sin\theta)t + \frac{1}{2}(-g)t^2 \tag{4}$$

[Using $s = ut + \frac{1}{2}at^2$, a is taken as $-g$ because the vertical component of the velocity decreases as the ball moves upwards]

Substituting Equation (3) into Equation (4):

$$y = (v\sin\theta)\left|\frac{x}{v\cos\theta}\right| + \frac{1}{2}(-g)\left|\frac{x}{v\cos\theta}\right|^2$$

$$\therefore \quad y = x\tan\theta - \left(\frac{g}{2v^2\cos^2\theta}\right)x^2 \qquad \left[\tan\theta = \frac{\sin\theta}{\cos\theta}\right]$$

This equation is a parabola of the form $y = ax - bx^2$

Example

A boy kicks a football such that the ball follows the path shown below (Figure 3.3.4).

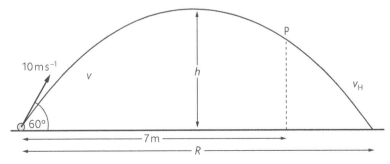

Figure 3.3.4

✅ *Exam tip*

When asked to find the velocity of the object at any point along the parabolic path, remember to find the horizontal and vertical components at that point first.

✅ *Exam tip*

Do not learn the formula for time of flight, horizontal range and maximum height. Just resolve the initial velocity into a horizontal and vertical component and apply the equations of motion.

Ignoring air resistance, calculate:

i the initial horizontal component of the velocity of the ball

ii the initial vertical component of the velocity of the ball

iii the maximum height h achieved by the ball

iv the time taken to reach the maximum height h

v the time taken to cover the distance R, where R is the range of the projectile

vi the distance R

vii the velocity of the ball at the point P, where P is 7 m from the initial starting point.

Sketch Figure 3.3.4 and label it A. Sketch the path B, such that the effect of air resistance on the ball is not ignored.

i Initial horizontal component $V_H = 10\cos 60° = 5\,\text{ms}^{-1}$.

ii Initial vertical component $V_V = 10\sin 60° = 8.66\,\text{ms}^{-1}$.

iii $u = 8.66\,\text{ms}^{-1}, v = 0\,\text{ms}^{-1}, a = -9.81\,\text{ms}^{-2}$

$$v = u^2 + 2as$$
$$0 = (8.66)^2 + 2(-9.81)s$$
$$2(-9.81)s = (8.66)^2$$
$$s = \frac{(8.66)^2}{2(9.81)} = 3.82\,\text{m}$$

Maximum height $h = 3.82\,\text{m}$

iv $u = 8.66\,\text{ms}^{-1}, v = 0\,\text{ms}^{-1}, a = -9.81\,\text{ms}^{-2}$

$$v = u + at$$
$$0 = 8.66 + (-9.81)t$$
$$t = \frac{8.66}{9.81} = 0.883\,\text{s}$$

Time taken to reach maximum height $= 0.883$ seconds

v Time taken to cover the horizontal distance:

$R = 2(0.883) = 1.766$ seconds

vi $u = 5\,\text{ms}^{-1}, a = 0\,\text{ms}^{-2}, t = 1.766\,\text{s}$

$$s = ut + \frac{1}{2}at^2$$
$$= (5 \times 1.766) + \frac{1}{2}(0)(1.766)^2$$
$$= 8.83\,\text{m}$$

vii In order to find the velocity of the projectile at P, we need to find the velocity in the vertical and horizontal direction at P. The velocity in the horizontal direction is constant, since there is no component of acceleration in that direction.

$$v_H = 5\,\text{ms}^{-1}$$

The time taken to reach the point P is determined as follows:

$$s = ut + \frac{1}{2}at^2$$
$$7 = 5t + \frac{1}{2}(0)t^2$$
$$t = \frac{7}{5} = 1.4\,\text{s}$$

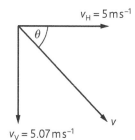

$v_H = 5\,\text{ms}^{-1}$

θ

v

$v_V = 5.07\,\text{ms}^{-1}$

Figure 3.3.5

The vertical component of velocity at time $t = 1.4$ seconds is determined as follows (Figure 3.3.5):

$$v = u + at$$
$$V_V = 8.66 + (-9.81)(1.4)$$
$$V_V = -5.07 \, \text{m s}^{-1}$$

The resultant velocity is determined as follows:

$$v = \sqrt{V_H^2 + V_V^2}$$
$$v = \sqrt{5^2 + 5.07^2}$$
$$v = 7.12 \, \text{m s}^{-1}$$
$$\theta = \tan^{-1}\left(\frac{V_V}{V_H}\right) = \tan^{-1}\left(\frac{5.07}{5}\right) = 45.3°$$

Figure 3.3.6 shows the effect of air resistance on the motion of the ball.

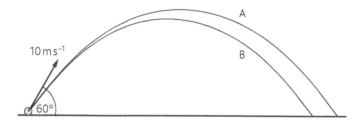

Figure 3.3.6 *Effect of air resistance on the motion of the ball*

Key points

- When an object is projected at an angle it follows a parabolic path.
- The object travels both horizontally and vertically simultaneously.
- The horizontal and vertical motions are treated separately.
- The horizontal component of the velocity remains constant if air resistance is ignored.
- Gravity acts vertically and affects the vertical component of the velocity.

4 Dynamics

4.1 Dynamics 1

Learning outcomes

On completion of this section, you should be able to:

■ define linear momentum

■ state Newton's first and second laws

■ define the newton.

Definition

Linear momentum is defined as the product of a body's mass and velocity in a given direction.

Equation

$p = mv$

p – momentum/kg m s^{-1}
m – mass of body/kg
v – velocity of body/m s^{-1}

Newton's first law

Newton's first law of motion states that a body stays at rest or if moving continues to move with uniform velocity unless acted upon by an external force.

Newton's second law

Newton's second law states that the rate of change of momentum is proportional to the applied force and takes place in the direction in which the force acts.

Linear momentum

The **linear momentum** (p) of an object is the product of its mass (m) and velocity (v). Momentum is a vector quantity. The direction of the momentum is in the same direction as the velocity (Figure 4.1.1). The unit of momentum, by definition is kg m s^{-1}.

Figure 4.1.1 Defining linear momentum

Newton's first law of motion

If a book is placed on a table it will stay there until a force is applied to it to make it move. If a rock is thrown in outer space, it will continue to move indefinitely in a straight line until the gravity of some object in space affects its motion. It is difficult to grasp the concept that an object can move in a straight line forever. On Earth frictional forces are always present. If a ball is kicked it will eventually come to rest because friction slows down the motion of the ball.

Newton stated that a body will stay at rest or if it is moving, will continue to move with a constant velocity unless acted upon by an external force. This law is stating that a force is required to produce a change in velocity (acceleration) (Figure 4.1.2).

A force is required to cause an object to accelerate from rest

A force is required to make an object accelerate when moving with a constant velocity

Figure 4.1.2

Newton's second law

Newton's second law states that the rate of change of momentum is proportional to the applied force and takes place in the direction in which the force acts.

Mathematically this law can be expressed as follows:

$$F \propto \frac{\Delta p}{\Delta t} \qquad (1)$$

Where F is force, Δp is the change in momentum and Δt is the change in time.

Consider an object of mass m travelling with a velocity u. A force F is applied to the object for t seconds and its velocity changes to v.

The initial momentum is mu.

The final momentum is mv.

The change in momentum is $mv - mu$.

$$\therefore \qquad \Delta p = mv - mu \qquad\qquad (2)$$

Substituting Equation (2) into Equation (1):

$$\therefore \qquad F \propto \frac{mv - mu}{t}$$

The proportionality sign is now replaced with an equal sign and a proportionality constant k is included.

$$F = k\left(\frac{mv - mu}{t}\right)$$

$$F = km\left(\frac{v - u}{t}\right)$$

But acceleration $a = \dfrac{v - u}{t}$

$$\therefore \qquad F = kma$$

1 newton is the force required to give a mass of 1 kg an acceleration of $1\,\mathrm{m\,s^{-2}}$. The reason why the newton is defined in this way is to make k in the equation equal to 1.

Example

A wooden block of mass 0.50 kg rests on a rough horizontal surface. A force of 15 N is applied to the block. The frictional force acting on the block is 6 N (Figure 4.1.3). Calculate the acceleration of the block.

$$\text{Acceleration } a = \frac{F}{m} = \frac{15 - 6}{0.5} = 18\,\mathrm{m\,s^{-2}}$$

Example

A box of mass 60 kg is being pulled along a rough surface as shown in Figure 4.1.4. Calculate:

a the component of the 80.2 N force in the OX direction

b the component of the 50 N force in the OX direction

c the acceleration of the box in the direction of OX if the frictional force acting on it is 25 N.

---

a Component in the OX direction $80.2 \times \cos 35° = 65.7\,\mathrm{N}$

b Component in the OX direction $50 \times \cos 67° = 19.5\,\mathrm{N}$

c Total horizontal force acting in the direction OX to the right

$$65.7 + 19.5 = 85.2\,\mathrm{N}$$

Resultant force acting on the box $= 85.2 - 25\,\mathrm{N} = 60.2\,\mathrm{N}$

$$a = \frac{F}{m} = \frac{60.2}{60} = 1.00\,\mathrm{m\,s^{-2}}$$

Key points

- Linear momentum is the product of a body's mass and velocity.

- Newton's first law states that a body stays at rest or if moving continues to move with uniform velocity unless acted upon by an external force.

- Newton's second law states that the rate of change of momentum is proportional to the applied force and takes place in the direction in which the force acts.

- 1 newton is the force required to give a mass of 1 kg an acceleration of $1\,\mathrm{m\,s^{-2}}$.

Equation

$$F = ma$$

F – force/N
m – mass/kg
a – acceleration/$\mathrm{m\,s^{-2}}$

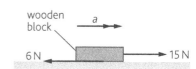

Figure 4.1.3

✓ Exam tip

Always remember that in the equation $F = ma$, F is resultant force.

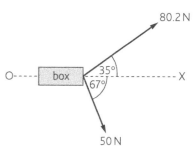

Figure 4.1.4

Impulse

Consider an object of mass m moving with a velocity u. A force F is applied to the object for a period of time t seconds. This force causes the velocity to increase to v. According to Newton's second law we can write

$$Ft = mv - mu$$

$mv - mu$ represents the change in momentum of the object and can be expressed as Δp, where p is momentum.

$$Ft = \Delta p$$

The quantity Ft is called **impulse**. The unit of impulse is the N s. The concept of impulse takes into account the time effect of a force. For example, in a game of cricket, the batsman can get the ball to go further when he strikes it with his bat, by keeping the bat in contact with the ball for a longer time. If the force exerted by the bat is exerted for a longer period of time, the ball will have a greater change in momentum.

If a fieldsman had to catch the ball he would let the ball fall into his hand, while at the same time moving his hand in the same direction the ball was travelling in. It takes a longer time for the momentum of the ball to reduce to zero. This reduces the force exerted on his hand by the ball.

F–t graphs

A force–time graph can be used to illustrate how a force varies over a period of time as it acts on an object. The first graph shows a constant force F being applied over a time interval t.

Suppose in a game of tennis, a player strikes the ball with his racket. The second graph shows how the force exerted by the racket on the ball would vary with time. In practice it is difficult for the player to apply a constant force to the ball. The ball deforms as it is struck by the racket and the force applied to it cannot be constant.

The area under a force–time graph represents the change in momentum of the object in question. For example the area under the graph represents the change in momentum of the tennis ball.

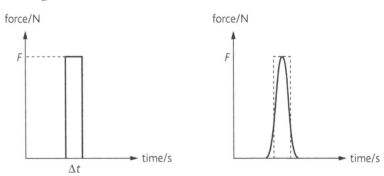

Figure 4.2.1 *Force–time graphs*

Mass and weight

The **mass** of a body is the amount of matter contained in it. Bodies have a property associated with them, called **inertia**. It is the reluctance of a body to start moving when it is at rest. It is also the reluctance of a body to stop moving, once it is in motion. Mass is a measure of a body's inertia.

Mass is a scalar quantity and the SI unit is the kilogram (kg). Suppose your mass is 60 kg on Earth. If you were to take a trip to the Moon, your mass will still be 60 kg.

The **weight** of an object is the force exerted by gravity on it. Weight is dependent on the gravitational field strength. Therefore, if you were to travel to the Moon, where the gravitational field strength is less than that of the Earth, you would have a much smaller weight than on the Earth. Weight is a vector quantity and the SI unit is the newton (N).

Newton's third law

For an object resting on the surface of a table, there are two forces acting on it. The weight of the object acts downwards. The object is not moving in a vertical direction. Therefore, there must be an equal force acting vertically upwards to balance the weight of the object. This force is called the **normal reaction** to the surface (Figure 4.2.2).

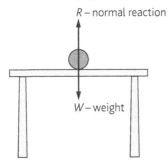

R – normal reaction

W – weight

Figure 4.2.2 *Forces acting on an object on a table*

Examples of Newton's third law of motion

1. Force between parallel current-carrying conductors

When two parallel current-carring conductors are adjacent to each other, a force is experienced. Consider two wires A and B parallel to each other, each carrying a current I in the same direction. Wire A exerts a force F_A on wire B. Wire B exerts a force F_B on wire A. F_A and F_B are of equal magnitude but act in opposite directions. When current in the wires flows in the same direction, the force between them is attractive. When the current in the wires flows in opposite directions, the force between them is repulsive (Figure 4.2.3).

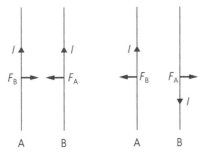

Figure 4.2.3 *Force between current-carrying conductors*

2. Force between charged objects

When two charges are placed close to each other, a force is experienced between them. Consider two charged objects A and B. Object A exerts a force F_A on object B. Object B exerts an equal and opposite force F_B on object A. If the charges are the same (both positive or both negative), the force is repulsive. If the charges are different, the force is attractive (Figure 4.2.4).

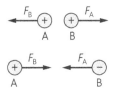

Figure 4.2.4 Force between charged objects

3. Gravitational force between two masses

The Moon orbits the Earth. The Earth's gravitational field provides the centripetal force required for the Moon to orbit the Earth. The Earth exerts a force F_E on the Moon. At the same time, the Moon exerts a force F_M on the Earth (Figure 4.2.5).

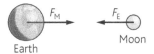

Figure 4.2.5 Force between two masses

4. A rocket

When a rocket is launched, a large volume of gas is expelled from the exhaust. As gas is being pushed out, it experiences a change in momentum, resulting in a downward force. According to Newton's third law, the gas exerts an equal and opposite force on the rocket, therefore causing it to accelerate.

5. A hovering helicopter

A helicopter is able to fly because of Newton's third law. The rotating blades of a helicopter exert a downward force on the air around them. The air has a change in momentum downwards, giving rise to a force. According to Newton's third law, the air exerts an equal and opposite force on the blades. If this force is equal to the weight of the helicopter, the resultant vertical motion is zero and the helicopter remains stationary. If the force is greater than the weight of the helicopter, it accelerates upwards.

When applying Newton's third law it is important to note the following:

- The two forces are at all times equal in magnitude.
- The two forces act in opposite directions.
- The two forces are of the same type (gravitational, electric, etc.)
- The two forces act on different bodies.

Example

A ball of mass $0.35\,kg$ hits a wall with a speed of $12\,m\,s^{-1}$ and rebounds from the wall along its initial path with a speed of $7.2\,m\,s^{-1}$. The impact with the wall lasts for $0.2\,s$. Calculate the average force exerted by the wall on the ball.

$$F = \frac{mv - mu}{t} = \frac{(0.35 \times -7.2) - (0.35 \times 12)}{0.2} = 33.6\,N$$

Example

Figure 4.2.6 below shows how the force acting on an object varies with time. The mass of the object is $2\,kg$ and is initially at rest.

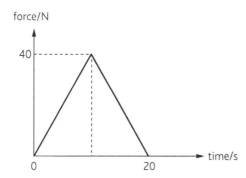

Figure 4.2.6

Calculate:

a the change in momentum of the mass during the first 10 seconds
b the velocity of the object after 10 seconds
c the acceleration of the object during the first 10 seconds
d the change in momentum of the mass during the 20-second interval.

a Change in momentum = area under graph
$$= \tfrac{1}{2} \times 10 \times 40 = 200\,N\,s$$

b $\Delta p = mv - mu$
$200 = 2(v) - 2(0)$
$v = 100\,m\,s^{-1}$

c Acceleration $= \dfrac{v - u}{t} = \dfrac{100 - 0}{10} = 10\,m\,s^{-2}$

d Change in momentum = total area under graph
$$= \tfrac{1}{2} \times 20 \times 40 = 400\,N\,s$$

Key points

- Impulse is the time effect of a force.
- The area under a force–time graph is equal to the change in momentum.
- Mass is the measure of the inertia of a body.
- Weight is the force exerted by gravity on it.
- The mass of a body remains fixed and is not affected by location.
- The weight of a body can vary if the gravitational field strength varies.
- Newton's third law states that if a body A exerts a force on body B, body B exerts an equal and opposite force on body A.

4.3 Collisions

Figure 4.3.1 *Before collision*

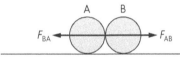

Figure 4.3.2 *During collision*

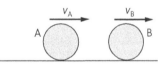

Figure 4.3.3 *After collision*

Definition

The principle of conservation of momentum states that for any system, the total momentum before collision is equal to the total momentum after collision provided that no external forces act on the system.

The principle of conservation of momentum

Figure 4.3.1 shows two identical balls A and B on a horizontal surface. Ball B is at rest and ball A is moving towards B with a velocity of v. The mass of each ball is m. Ball A eventually collides with ball B.

During the collision of the balls, the magnitude of the force that ball A exerts on ball B is F_{AB} and the magnitude of the force that ball B exerts on ball A is F_{BA} (Figure 4.3.2).

The balls are in contact for a length of time Δt. After the collision, the speed of ball A is v_A, and the speed of ball B is v_B in the directions shown in Figure 4.3.3.

Analysis

Ball B

The initial momentum of ball B is zero because it is initially at rest.

The final momentum of ball B is mv_B.

The change in momentum of the ball B is therefore mv_B.

According to Newton's second law

$$F_{AB}\Delta t = mv_B$$

Ball A

The initial momentum of ball A is mv

The final momentum of ball A is mv_A.

The change in momentum of the ball A is therefore $mv_A - mv$.

According to **Newton's second law**

$$F_{BA}\Delta t = mv_A - mv$$

Total momentum

Total momentum before the collision = mv

Total momentum after the collision = $mv_A + mv_B$

According to **Newton's third law**, the magnitude of the force exerted on ball B by ball A (F_{AB}) is equal to the magnitude of the force exerted by ball A by ball B (F_{BA}). They act in opposite directions.

$$\therefore \qquad F_{AB} = -F_{AB} \qquad (1)$$

(Multiplying Equation (1) by Δt) $\quad F_{AB}\Delta t = -F_{AB}\Delta t$

$$mv_B = -(mv_A - mv)$$

$$mv_B = mv - mv_A$$

$$\therefore \qquad mv = mv_A + mv_B \qquad (2)$$

Equation (2) shows that the total momentum just before the collision is equal to the total momentum just after the collision. This example illustrates the principle of conservation of momentum.

Momentum is conserved in all collisions.

Elastic and inelastic collisions

In all collisions, momentum is conserved. This is not the case with kinetic energy. Kinetic energy is the energy possessed by a body by virtue of its motion. If kinetic energy is conserved in a collision, it is said to be an **elastic collision**. If kinetic energy is not conserved, the collision is said to be **inelastic**.

Table 4.3.1 compares elastic and inelastic collisions.

Example

An object of mass 3 kg travelling at $6\,\text{m s}^{-1}$ strikes another object of mass 5 kg travelling at $1\,\text{m s}^{-1}$ in the same direction. The objects stick together and move off with a velocity v. Calculate v (Figure 4.3.4).

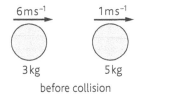 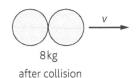

Figure 4.3.4

Assuming no external forces act on the system (the two objects):

Total momentum before collision = total momentum after collision

$$(3 \times 6) + (5 \times 1) = (3 + 5)v$$

$$23 = 8v$$

$$v = \frac{23}{8} = 2.88\,\text{m s}^{-1}$$

Both objects move off with a velocity of $2.88\,\text{m s}^{-1}$ in the same direction both objects were originally moving in.

Always assume a sign convention.

- Motion to the right is taken as positive velocities.
- Motion to the left is taken as negative velocities.

Example

A stationary nucleus of mass $3.65 \times 10^{-25}\,\text{kg}$ decays to produce two particles A and B. The particles A and B move off in opposite directions.

Mass of A $= 6.64 \times 10^{-27}\,\text{kg}$

Mass of B $= 3.59 \times 10^{-25}\,\text{kg}$

The initial speed of A is $1.7 \times 10^7\,\text{m s}^{-1}$. Calculate the initial speed of B (Figure 4.3.5).

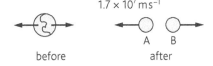

Figure 4.3.5

Total momentum before = total momentum after

$$0 = (6.64 \times 10^{-27} \times -1.7 \times 10^7) + (3.59 \times 10^{-25} \times v)$$

$$0 = -1.13 \times 10^{-19} + 3.59 \times 10^{-25}v$$

$$v = \frac{1.13 \times 10^{-19}}{3.59 \times 10^{-25}} = 3.14 \times 10^5\,\text{m s}^{-1}$$

Definitions

An elastic collision is one in which kinetic energy is conserved.

An inelastic collision is one in which kinetic energy is not conserved.

Table 4.3.1

	Elastic collision	Inelastic collision
Momentum conserved	Yes	Yes
Kinetic energy conserved	Yes	No
Total energy conserved	Yes	Yes

Key points

- The principle of conservation of momentum states that for a system, the total momentum before collision is equal to the momentum after collision provided no external forces act on the system.

- In an elastic collision, kinetic energy is conserved.

- In an inelastic collision, kinetic energy is not conserved.

- Total energy is conserved in all collisions.

Answers to questions that require calculation can be found on the accompanying CD.

1 When a substance is heated, the energy required to raise its temperature is given by the expression:

Energy required $=$ mass $\times c \times$ change in temperature

where c is a constant.

  a State the quantities that are SI base quantities. [2]

  b Determine the units of the following, in terms of SI base units:

    i energy required [2]

    ii the constant c. [2]

2 State three base quantities and their corresponding base units. [3]

3 A small metal sphere of radius r is dropped into a viscous fluid. As it falls at a speed v, it experiences a drag force F, where $F = krv$. k is a constant. State the SI base units of k. [3]

4 The frictional force acting on a glass marble falling through a viscous fluid is given by $F = 6\pi r\eta v$, where r is the radius of the marble, η is the viscosity of the fluid and v is the velocity of the marble. Determine SI units for viscosity. [3]

5 With the aid of an example, explain what is meant by the statement "The magnitude of a physical quantity is written as the product of a number and a unit". [2]

6 Explain why an equation must be homogenous with respect to the units if it is to be correct. [1]

7 Write an equation which is homogenous, but incorrect. [2]

8 a Explain the principle which underlies the checking of the balance of equations using base units. [2]

  b State one limitation of using base units to balance equations. [1]

9 a State the difference between vector and scalar quantities. [2]

  b Give two examples of a vector quantity and two examples of a scalar quantity. [4]

10 An object is being pulled with force of 5 N acting at an angle of 30° to the horizontal. Calculate the horizontal and vertical components of the 5 N force. [2]

11 An object is acted upon by two forces P and Q. The two forces act at an angle of θ. Given that $P = 6$ N and $Q = 8$ N, determine the resultant force acting on the object when:

  a $\theta = 0°$ b $\theta = 180°$

  c $\theta = 90°$ d $\theta = 120°$. [8]

12 Distinguish between a systematic and a random error. [3]

13 A student wishes to measure the diameter of a piece of wire using a micrometre screw gauge.

  a How can the student eliminate any systematic error in the measurement? [2]

  b How can he reduce the random error in measuring the diameter? [2]

14 Distinguish between precision and accuracy. [3]

15 A quantity S is determined from the equation $S = P - Q$.

$P = 6.12 \pm 0.02$ m and $Q = 1.84 \pm 0.02$ m. Calculate the percentage uncertainty in S. [2]

16 The resistance of piece of wire is given by $R = V/I$. A student wishes to determine the resistance of a piece of wire. V and I are measured.

$V = 12 \pm 0.4$ V

$I = 1.0 \pm 0.2$ A

Calculate the resistance of the wire and include the absolute uncertainty R. [3]

17 The mean diameter of a piece of wire is 0.6 ± 0.02 mm. Calculate the percentage uncertainty in:

  a the diameter [2]

  b the cross-sectional area of the wire. [3]

18 The volume of a cylinder is given by the expression $V = \pi r^2 h$. The volume and height of a cylinder is measured as:

$V = 12.0 \pm 0.5$ cm³

$h = 21.0 \pm 0.1$ cm

Calculate the radius of the cylinder, with its uncertainty. [5]

19 a Differentiate between displacement and distance travelled. [2]

b State how you would determine velocity from a displacement–time graph. [1]

c State how you would determine the displacement and acceleration from a velocity–time graph. [2]

20 Define the following terms:

a velocity [2]

b acceleration. [2]

21 The variation with time t of the velocity v of a cyclist travelling down a slope is illustrated below.

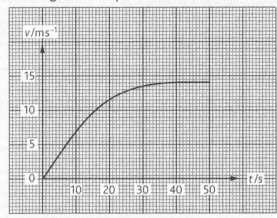

The cyclist reaches a constant velocity after 40 s. Using the graph estimate:

a the maximum velocity of the cyclist [1]

b the initial acceleration of the cyclist [3]

c the total distance travelled before reaching constant velocity. [3]

22 A car approaches a traffic light with a speed of 20 m s^{-1}. The light changes to red. The driver applies the brake when at a distance of 40 m from the lights. Calculate the deceleration of the car if it comes to rest at the lights. [3]

23 At a sports day at school, Akil runs a 100 m race. He accelerates from the blocks at a rate of 2 m s^{-2} for 4.5 seconds. He runs the remainder of the race at a constant speed.

a Calculate his speed after the first 5 s. [2]

b Calculate the distance travelled during the first 5 s. [2]

c Determine the time taken for the race. [3]

d Sketch the velocity–time graph for the race. [3]

24 A metal ball is thrown vertically upwards with an initial velocity of 15 m s^{-1}.

a Ignoring air resistance, determine:

i the distance travelled by the ball after 0.9 seconds [2]

ii the velocity of the ball after 0.9 seconds [2]

iii the time taken for the ball to reach its maximum height [2]

iv the maximum height reached by the ball. [2]

b The ball falls to the ground and bounces twice. Sketch the velocity–time graph for the metal ball. [3]

c Show how it is possible to determine the distance travelled by the ball between the first bounce and the second bounce. [2]

25 An object is projected with an initial velocity of 5 m s^{-1} at an angle of 30° to the horizontal.

Ignoring air resistance:

i Calculate the initial horizontal component of the velocity of the object. [1]

ii Calculate the initial vertical component of the velocity of the object. [1]

iii Sketch a graph to show the variation of the horizontal component with time. [1]

iv On the same graph, show the variation of the vertical component with time. [2]

v Calculate the maximum height h achieved by the object. [2]

vi Calculate the time taken to reach the maximum height h. [2]

vii Calculate the time taken to cover the distance R, where R is the range of the projectile. [2]

viii Calculate the range R. [2]

ix Calculate the velocity of the ball at the point P, where P is 2 m from the initial starting point. [3]

26 A box of mass 25 kg is pulled up a smooth inclined plane at 40° to the horizontal by a rope which is parallel to the plane.

a Sketch a diagram to show the forces acting on the box. [3]

b Calculate the component of the weight of the box acting parallel to the inclined plane. [2]

c The tension in the rope is 250 N. Calculate the acceleration of the box. [3]

d Calculate the reaction force between the box and the plane. [2]

5.1 Archimedes' principle, friction and terminal velocity

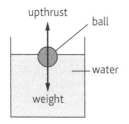

Figure 5.1.1 *The forces acting on an object immersed in a fluid*

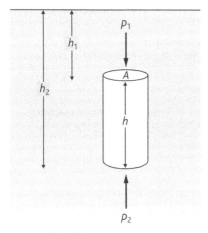

Figure 5.1.2 *Origin of an upthrust*

Definition

Archimedes' principle states that when a body is totally or partially submerged in a fluid, it experiences an upthrust which is equal to the weight of the fluid displaced.

Archimedes' principle

When a beach ball is placed in water there are two forces acting on it.

1 The weight of the ball, which acts vertically downwards.
2 The **upthrust**, which is a force acting vertically upwards.

When the ball is placed in water, it displaces some water. The weight of this displaced water is equal to the upthrust (Figure 5.1.1).

Origin of the upthrust

Consider a cylinder of height h and cross sectional area A at a distance of h_1 below the surface of a fluid of density ρ (Figure 5.1.2).

(See 19.1 and 19.2 for coverage of density and pressure.)

$$\text{Volume of fluid displaced} = \text{volume of cylinder} = Ah$$
$$\text{Mass of fluid displaced} = \text{density} \times \text{volume} = \rho Ah$$
$$\text{Weight of fluid displaced} = \text{mass} \times \text{gravitational field strength}$$
$$= \rho Ahg \tag{1}$$

$$\text{Pressure} \qquad p_1 = \rho gh_1$$
$$\text{Force exerted on top of surface} \quad F_1 = pA = \rho gh_1 A$$
$$\text{Pressure} \qquad p_2 = \rho gh_2$$
$$\text{Force exerted on bottom surface} \quad F_2 = pA = \rho gh_2 A$$
$$\text{Upthrust} = F_2 - F_1$$
$$= \rho gh_2 A - \rho gh_1 A$$
$$= \rho gA(h_2 - h_1)$$
$$\text{But} \qquad h = h_2 - h_1$$
$$\text{Upthrust} = \rho gAh \tag{2}$$

From Equations (1) and (2), upthrust = weight of fluid displaced.

Resistive forces

When a ball is rolled on the floor, it eventually comes to rest. The reason why the ball comes to rest is because of friction. **Friction** is a force that opposes the motion of an object. Frictional forces:

1 slow down the motion of moving objects and
2 prevent movement between two stationary objects in contact with each other.

Heat is produced when work is done against friction.

In machines, there are frictional forces acting on moving parts. The frictional forces cause thermal energy to be produced. This makes machines less efficient. Lubricating moving parts with oil or grease,

inside the machine helps reduce friction. This reduces the amount of wear and tear inside the machine and makes it more efficient.

Friction arises because surfaces are not completely smooth. Even though they may appear that way, at the microscopic level they are actually jagged and rough.

Frictional forces can be reduced by:

- using lubricants such as oil or grease
- using rollers and ball bearings between surfaces in contact
- polishing surfaces to ensure they are smooth as possible.

Drag force is the force that opposes the motion of an object as it moves through a fluid. **Air resistance** is a special type of frictional force which acts on objects as they travel through air. At low velocities, the air resistance is proportional to the velocity of the object ($F = kv$). At higher velocities, the air resistance is proportional to the square of the velocity of the object ($F = kv^2$).

Terminal velocity

Consider a parachutist jumping from an aircraft. When the parachutist jumps from an aircraft, his initial velocity is zero. The initial acceleration is $9.81\,\mathrm{m\,s^{-2}}$. The initial force acting on the parachutist is his weight, which acts downwards. There is no drag force acting on him at the start because the initial velocity is zero. As he falls, his velocity increases and his acceleration decreases. The drag force acting on him is proportional to his velocity and acts upwards. The drag force therefore increases as he falls. The resultant force acting on him is $F_R = W - D$, where W is the weight of the parachutist and D is the drag force acting on him. At some point in the fall, his weight becomes equal to drag force. At this point, the resultant force acting on him is zero. His acceleration is also zero, which means that he is falling at a constant velocity. He has now reached **terminal velocity** (Figure 5.1.3).

Example

An object has a mass of 2.2 kg. When the object falls in air, the air resistance F is given by $F = kv^2$, where v is the velocity of the object and $k = 0.039\,\mathrm{N\,s^2\,m^{-2}}$.

Calculate:

a the weight of the object

b the terminal velocity of the object

c the acceleration of the object when it is falling with a velocity of $10\,\mathrm{m\,s^{-1}}$.

a $W = mg = 2.2 \times 9.81 = 21.6\,\mathrm{N}$

b At terminal velocity, the resultant force acting on the object is zero.

$$\therefore \quad W = F$$
$$F = kv^2$$
$$21.6 = (0.039)v^2$$
$$v = \sqrt{\frac{21.6}{0.039}} = 23.5\,\mathrm{m\,s^{-1}}$$

c When $v = 10\,\mathrm{m\,s^{-1}}$, air resistance $F = (0.039)(10)^2 = 3.9\,\mathrm{N}$

Resultant force acting on object $= W - F = 21.6 - 3.9 = 17.7\,\mathrm{N}$

$$a = \frac{F}{m} = \frac{17.7}{2.2} = 8.04\,\mathrm{m\,s^{-2}}$$

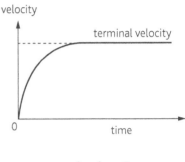

Figure 5.1.3 *Terminal velocity*

Key points

- Archimedes' principle states that when a body is totally or partially submerged in a fluid, it experiences an upthrust which is equal to the weight of the fluid displaced.

- Friction opposes motion.

- When work is done against friction, heat is produced.

- An object reaches terminal velocity when it falls through a fluid.

- At terminal velocity, the weight of the object is equal to the drag force. The resultant force is zero and the acceleration is zero.

Polygon of forces

Suppose three forces P, Q and R act on an object. Figure 5.2.1(a) shows the free body diagram. A **free body diagram** shows all the forces acting on it.

The object is in equilibrium. The forces can be used to draw a **vector triangle** (Figure 5.2.1(b)). The sides of the triangle represent the magnitude of the forces. Since the object is in equilibrium, the triangle drawn will be a closed triangle. If the object was not in equilibrium, the vector triangle will not be closed. The vector triangle is drawn as follows:

1 One of the forces acting on the object is selected and drawn first (e.g. P).

2 Moving in an anticlockwise direction, the next force is drawn. The force R is drawn by starting from the arrowhead of P.

3 The force Q is then drawn by starting at the arrowhead of R.

The directions of all the forces in the vector triangle are exactly the same as in the free body diagram.

Suppose an object is in equilibrium when acted upon by four forces, P, Q, R and S. In this case a vector polygon can be drawn. Using the same principles used for the vector triangle, the polygon can be drawn (Figure 5.2.2).

Suppose two forces P and Q act on an object. The object is not in equilibrium. The resultant force acting on the object is R. In order for the object to be in equilibrium, a force of equal magnitude to R, acting in a direction opposite to R must be exerted on the body (Figure 5.2.3).

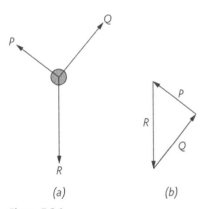

Figure 5.2.1
(a) Three forces acting on a body
(b) A vector triangle

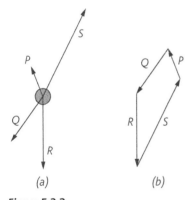

Figure 5.2.2
(a) The free body diagram
(b) The vector polygon

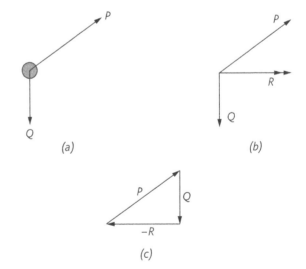

Figure 5.2.3 (a) Free body diagram (b) Resultant force (c) The vector triangle

Example

A particle of mass $0.51\,\text{kg}$ is supported by a string attached to a fixed point. It is being pulled by a horizontal force of $3.2\,\text{N}$.

a Sketch a diagram to show the forces acting on the particle.

b Draw a vector triangle and hence calculate the tension in the string.

c Calculate the angle that the string makes with the vertical.

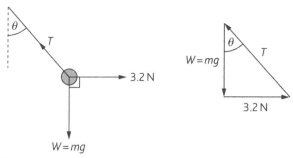

Figure 5.2.4

$W = mg = 0.51 \times 9.81 = 5.0\,\text{N}$

Using Pythagoras' theorem

$$T^2 = 5^2 + 3.2^2$$
$$T = \sqrt{35.24} = 5.9\,\text{N}$$
$$\tan\theta = \frac{3.2}{5.0}$$
$$\theta = \tan^{-1}\left|\frac{3.2}{5.0}\right| = 32.6° \quad \text{(Figure 5.2.4)}$$

Centre of gravity

Finding the centre of gravity of an irregular-shaped lamina

A lamina is a thin sheet of stiff material. In order to locate the centre of gravity of an irregular-shaped lamina, the following steps are taken:

1 Two small holes are made near the edge of the lamina.

2 A nail is placed through one of the holes and the lamina is made to hang freely from it.

3 A string with a mass attached to it is then attached to the nail.

4 A pencil is used to mark several points on the lamina where the string hangs.

5 The mass and string is removed and a straight line is drawn through the points made in step 4.

6 A nail is placed through the second hole and the lamina is made to hang freely.

7 The string with a mass attached to it is then attached to the nail.

8 A pencil is used to mark several points on the lamina where the string hangs.

9 The mass and string is removed and a straight line is drawn through the points made in step 8.

10 The point of intersection of the two lines drawn is the centre of gravity of the lamina (Figure 5.2.5).

Key points

- A free body diagram shows all the forces acting on a body.

- For an object in equilibrium when several forces act on it, a closed vector polygon can be drawn.

- The centre of gravity of a body is the point through which all the weight of a body seems to act.

Definition

The **centre of gravity** of a body is the point through which all the weight of a body appears to act.

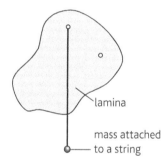

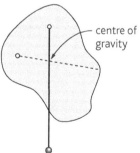

Figure 5.2.5 *Finding the centre of gravity of an irregular shaped lamina*

Definition

The moment of a force is defined as the product of the force and the perpendicular distance of the line of action of the force from the pivot.

The moment of a force

There is a reason why a door knob is positioned at the edge of a door. It is not positioned at the middle of the door where a larger force would be required to open the door. It is positioned at the edge the door in order for there to be a large **moment** about the hinges of the door. The moment of a force about a pivot is the turning effect of a force. The moment of a force is defined as the product of the force and the perpendicular distance of the line of action of the force from the pivot. The SI unit is the newton metre (Nm).

Consider Figure 5.3.1.

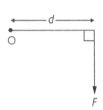   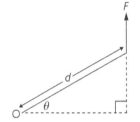

Figure 5.3.1 Defining moment of a force   *Figure 5.3.2*

The moment of force F about $O = F \times d$

Always remember, when calculating the moment of a force, the distance used is at right angles to the line of action of the force from the pivot. In Figure 5.3.2, the moment of the force F determined as follows:

The moment of force F about $O = F \times d \cos\theta$

The principle of moments

Consider a plank P balancing on a pivot O. Forces F_1, F_2 and F_3 act on the plank as shown in Figure 5.3.3.

The force F_1 produces an anticlockwise moment $F_1 d_1$ about O.

The force F_2 produces an anticlockwise moment $F_2 d_2$ about O.

The force F_3 produces a clockwise moment $F_3 d_3$ about O.

Therefore, for the plank P to be in equilibrium

$$\text{Sum of clockwise moments} = \text{Sum of anticlockwise moments}$$
$$F_3 d_3 = F_1 d_1 + F_2 d_2$$

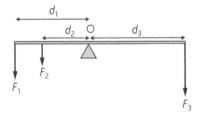

Figure 5.3.3 Applying the principle of moments

Definition

The **principle of moments** states that for a body to be in equilibrium, the sum of the clockwise moments must be equal to the sum of the anticlockwise moments about the same pivot.

Example

Consider a wheelbarrow filled with some sand. A construction worker is about to lift the wheelbarrow with a force P (Figure 5.3.4).

Calculate:

i   the minimum value of the vertical force P, needed to raise the legs of the wheelbarrow off the ground

ii   the magnitude of R when the legs of the wheelbarrow just leave the ground.

i   Taking moments about the centre of the wheel.

Sum of clockwise moments = Sum of anticlockwise moments

$$480 \times 0.6 = P \times 1.4$$
$$P = \frac{480 \times 0.6}{1.4}$$
$$= 206\,\text{N}$$

ii Just as the wheelbarrow is about to leave the ground, it will be in equilibrium.

Sum of upward forces = Sum of downward forces
$$R + 206 = 480$$
$$R = 480 - 206$$
$$R = 274\,\text{N}$$

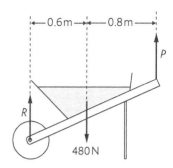

Figure 5.3.4

The torque of a couple

A **couple** consists of two equal and opposite forces whose lines of action do not coincide. A couple tends to produce rotation only. Consider two forces F_1 and F_2 acting on a steering wheel of a car (Figure 5.3.5).

The forces F_1 and F_2 are equal and have a turning effect or moment called a **torque**.

Torque of couple $=F_1 d$ or $F_2 d$

The SI unit of torque is the newton metre (Nm).

It should be noted that the resultant turning effect is not zero. The steering wheel rotates anticlockwise. The steering wheel is not in equilibrium when subjected to forces F_1 and F_2 only.

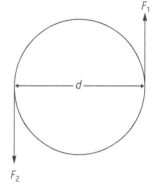

Figure 5.3.5 The torque of a couple

Example

A ruler of length 0.5 m is pivoted at its centre. Calculate the magnitude of the torque of the couple when equal and opposite forces of magnitude 3 N are applied as shown in Figure 5.3.6.

Torque of couple $= F \times d = 3 \times 0.5 \sin 60^\circ = 1.3\,\text{Nm}$

Conditions for equilibrium

In order for a system to be in equilibrium, the following conditions must apply.

- The resultant force acting on the system is zero.
- The resultant torque is zero.
- The resultant moment is zero.

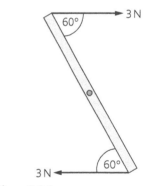

Figure 5.3.6

Key points

- The moment of a force is the product of the force and the perpendicular distance of the line of action of the force from the pivot.
- The principle of moments states that for a body to be in equilibrium, the sum of the clockwise moments is equal to the sum of the anticlockwise moments.
- A couple consists of two equal and opposite forces whose lines of action do not coincide.
- A couple tends to produce rotation only.
- The torque of a couple is the product of one of the forces and the perpendicular distance between the forces.

6.1 Work and energy

Energy

Energy is the capacity or ability to do work.

Various forms of energy include:

- mechanical (kinetic and potential)
- thermal
- chemical
- electrical
- nuclear
- solar (Table 6.1.1).

Energy conversion

The Sun is the primary source of energy for the Earth. Most of the energy reaching the Earth is in the form of light and infrared radiation (heat energy). Energy cannot be created but can be converted from one form to another. For example, gasoline has chemical energy locked up inside it. When burnt, the chemical energy converts to thermal and mechanical energy.

Table 6.1.1 *Examples of energy conversion*

	Example	Energy conversion
1	An incandescent bulb being switched on	Electrical energy to light and heat
2	A book falling from a shelf to the floor	Gravitational potential energy to kinetic energy and sound energy
3	A photovoltaic cell	Light energy into electrical energy
4	A battery	Chemical energy into electrical energy
5	A hydroelectric plant	Gravitational potential energy into kinetic energy and then into electrical energy
6	A lighted candle	Chemical energy into light and heat
7	A catapult being released	Elastic potential energy into kinetic energy

Energy conservation

Renewable sources of energy are derived from natural sources (sunlight, waves, wind, geothermal, hydroelectric) and are replenished over time.

Most of the energy used in homes, factories and transportation comes from fossil fuels. Fossil fuels are **non-renewable**. This means that they cannot be replenished. The Caribbean depends heavily on fossil fuels, such as oil and gas. As global reserves of oil and gas diminish, prices of these fuels increase. Therefore, in the Caribbean, there is a greater need for energy conservation.

To conserve energy:

- Switch off lights when leaving a room.
- Use fluorescent bulbs instead of incandescent bulbs.
- Use natural lighting.
- Do not leave refrigerator doors open.
- Switch off electrical appliances when not in use.
- Car pool with other people.
- Walk or use a bicycle instead of a car.

Alternative sources of energy in the Caribbean

Table 6.1.2

	Alternative source of energy	Main feature/use	Advantages	Disadvantages	Region for potential use
1	Solar	Solar energy harnessed from the Sun.	Abundance of sunlight in the Caribbean.	Sunlight varies throughout the day so insulated storage tanks are required.	All Caribbean territories
		Solar water heaters – homes and hotels.	Can be attached to the roof of existing buildings.	For large amounts of electricity many solar panels are required.	
				Batteries are required to store energy.	
		Solar panels that convert sunlight into electricity. Solar driers – used to dry crops.	Direct conversion of sunlight into electricity. Very effective at drying crops.		
2	Wind	Kinetic energy of wind converted into electrical energy using wind turbines.	Efficient method of converting wind into electricity.	Large capital cost. Affects environment.	Cuba Jamaica Barbados
				Wind is seasonal and variable.	
				Batteries are required to store energy.	
3	Hydroelectric	Gravitational potential energy of water stored in dams flow through turbines to produce electricity.	Efficient, reliable method of producing electricity.	Huge capital cost. Affects ecology. Problems with flooding.	Dominica Guyana (Amaila falls)
4	Geothermal	Thermal energy inside the Earth is used to produce steam to generate electricity.	Small land area required.	Very site-specific and expensive. Harmful gases may come up from the ground.	Guadelope St Lucia Dominica
5	Biofuels				All Caribbean territories
	▪ Biogas	Produced by bacteria breaking down plant and animal waste. Main constituent – methane.	Gas can be used for cooking and heating. One way of getting rid of waste material.	Greenhouse gases produced. Agricultural land is used to plant crops for fuel instead of food for consumption.	
	▪ Gasohol	Mixture of gasoline and alcohol.	Used as a fuel in some cars.		
	▪ Biodiesel	Made by chemically reacting vegetable oil with an alcohol.	Used as a fuel in diesel engines.	Some of the feedstock used for biodiesel is also used for food.	

Work

The **work done** by a force is the product of the force and the distance moved in the direction of the force.

$$W = Fs \qquad \text{(Figure 6.1.1)}$$
$$W = Fs\cos\theta \qquad \text{(Figure 6.1.2)}$$

The SI unit of work is the joule (J). 1 joule is the work done by a force of 1 N when it moves through a distance of 1 m in the direction of the force.

$$1\,J = 1\,Nm$$

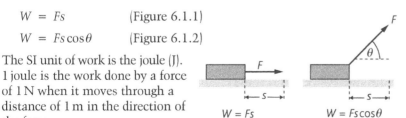

$$W = Fs \qquad\qquad W = Fs\cos\theta$$
Figure 6.1.1   **Figure 6.1.2**

When someone lifts an object, work is done by the upward force of the hand. If the object is held stationary in its final position, no work is being done by the upward force of the hand because it is stationary. However, the muscles in the arm get tired even though no work is being done.

Key points

- Energy is the capacity or ability to do work.

- The principle of conservation of energy states that energy can neither be created nor destroyed, but can be converted from one form to another.

- The work done by a force is the product of the force and the distance moved in the direction of the force.

Definition

The **kinetic energy** of a body is the energy possessed by virtue of its motion.

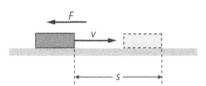

Figure 6.2.1 *Deriving the formula for kinetic energy*

Definition

The **potential energy** of a body is the energy possessed by it by virtue of its state or position.

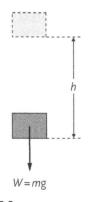

Figure 6.2.2

Kinetic energy

A cricket ball travelling through the air possesses kinetic energy. If the ball strikes a glass window it will break the glass. The energy used to break the glass comes from the kinetic energy possessed by the ball.

An object of mass m moving with a velocity v has a kinetic energy of $E_K = \frac{1}{2}mv^2$.

Deriving the equation for the kinetic energy of a body

Consider an object of mass m, travelling with a velocity of v. A constant force F acts on the object and brings it to rest while travelling through a distance s. The object decelerates at a rate of a (Figure 6.2.1).

Work done by the force F is $W = Fs$.

Using Newton's second law $(F = ma)$

$$\therefore \qquad W = mas$$

Considering the motion of the object:

$$\text{Initial velocity} = v \qquad \text{Final velocity} = 0 \qquad \text{Acceleration} = -a$$

$$v^2 = u^2 + 2as$$
$$0^2 = v^2 - 2as$$
$$v^2 = 2as$$
$$\therefore \qquad as = \frac{1}{2}v^2$$

Work done by the force F is $W = m\left(\frac{1}{2}v^2\right)$

The loss in kinetic energy of the object is equal to the work done by the force.

Therefore, the kinetic energy of the object is $E_K = \frac{1}{2}mv^2$

Potential energy

Potential energy can be classified as follows:

- Gravitation potential energy – The energy of a body by virtue of its position in a gravitational field.
- Electrical potential energy – The energy possessed by a charged body due to its position in an electric field.
- Elastic potential energy – The energy possessed by a body when deformed. (Example – a stretched spring has elastic potential energy.)

Deriving the equation for the change in gravitational potential energy of a body

Consider an object of mass m at a height above the ground. It moves vertically upwards with a constant velocity v and travels through a distance of h. In order to maintain a constant velocity, the upward force acting on the object must be equal to the weight of the object (Figure 6.2.2).

$$\text{Upward force} = F$$
$$\text{Downward force (Weight)} \quad W = mg$$
$$\therefore \qquad F = mg$$

Work done by force $= F \times d = mgh$

The work done by the force is equal to the gain in gravitational potential energy of the object.

$$\therefore \qquad E_p = mgh$$

Power and efficiency

Power is defined as the rate at which work is being done. The SI unit of power is the watt (W).

The work done by a force is defined as the product of the force and the distance moved in the direction of the force. $W = F \times d$

Power is defined as the rate at which work is being done. $P = \dfrac{W}{t}$

Therefore, we can write $P = \dfrac{F \times d}{t}$. But recall that $v = \dfrac{d}{t}$

$$\therefore \qquad P = Fv$$

The **efficiency** of a machine is defined as the ratio of the useful power output to the power input. It is expressed as a percentage.

Machines are not 100% efficient. There are energy losses present in machines. Friction between moving parts generate unwanted heat. Oiling and greasing moving parts reduces friction and increases the efficiency of machines.

Example

A cyclist pedalling along a horizontal road provides a power of $210\,\text{W}$ and reaches a steady speed of $6.2\,\text{m s}^{-1}$. The combined mass of the cyclist and bicycle is $112\,\text{kg}$.

a   Calculate:
  i   the kinetic energy of the cyclist and bicycle
  ii   the total resistive force acting on the forward motion.
b   The cyclist stops pedalling and allows the bicycle to come to rest. Assuming that the resistive force remains constant, calculate the distance travelled by the cyclist before coming to rest.
c   The cyclist decides to go up a slope. The angle of the slope is $30°$ to the horizontal. In order to maintain a constant speed of $6.2\,\text{m s}^{-1}$ up the slope, the cyclist pedals harder and supplies more power to the bicycle. Calculate this power.

a   i $E_K = \frac{1}{2}mv^2 = \frac{1}{2}(112)(6.2)^2 = 2.15 \times 10^3\,\text{J}$

  ii $P = Fv$
      $F = \dfrac{P}{v} = \dfrac{210}{6.2} = 33.9\,\text{N}$

b   Loss in kinetic energy = work done by resistive force
      $2.15 \times 10^3 = 33.9 \times d$
      $d = \dfrac{2.15 \times 10^3}{33.9} = 63.5\,\text{m}$

c   Component of the weight of the cyclist and bicycle down the slope is:
      $mg\sin\theta = 112 \times 9.81 \times \sin 30° = 549\,\text{N}$
      Frictional force $= 33.9\,\text{N}$

  Therefore, the cyclist must supply power to provide a force of $549 + 33.9 = 582.9\,\text{N}$ up the incline to maintain a speed of $6.2\,\text{m s}^{-1}$.
      $\therefore \quad P = Fv = 582.9 \times 6.2 = 3.61 \times 10^3\,\text{W}$ (Figure 6.2.3)

Equation

Efficiency $(\eta) = \dfrac{P_o}{P_i} \times 100\%$

P_o – useful power output/W
P_i – input power/W

Equation

$P = \dfrac{W}{t}$

P – power/W
W – energy/J
t – time/s

Equation

$P = Fv$

P – power/W
F – force/N
v – velocity/m s^{-1}

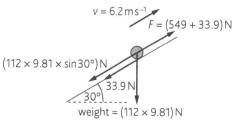

$v = 6.2\,\text{m s}^{-1}$
$F = (549 + 33.9)\,\text{N}$
$(112 \times 9.81 \times \sin 30°)\,\text{N}$
$33.9\,\text{N}$
$30°$
weight $= (112 \times 9.81)\,\text{N}$

Figure 6.2.3

Key points

■   The kinetic energy of a body is the energy possessed by virtue of its motion.

■   The gravitational potential energy of a body is the energy possessed by virtue of its position.

■   Power is defined as the rate at which work is being done.

■   Efficiency is defined as the ratio of the useful power output to the power input.

Revision questions 2

Answers to questions that require calculation can be found on the accompanying CD.

1   A small toy of mass 50 g attached to a string hangs from the roof of the inside of a car. The car accelerates horizontally and the string attached to the toy takes up a steady position at an angle of 25° to the vertical.

   a   Sketch a diagram to show the forces acting on the toy and indicate the direction of the acceleration of the car. [2]

   b   Calculate the magnitude of the resultant force acting on the toy. [3]

   c   Calculate the acceleration of the toy. [2]

2   a   State Newton's first and second laws of motion. [4]

   b   Using Newton's laws of motion, explain how a helicopter is able to hover above the ground. [4]

3   a   Define linear momentum. [2]

   b   State the SI unit of linear momentum. [1]

   c   An object of mass 0.6 kg is travelling with a velocity of 2.5 m s^{-1}.
     Calculate the kinetic energy and the momentum of the object. [3]
     Explain why a direction is required for one quantity and not the other. [2]

4   A cricketer throws a ball of mass 0.15 kg. The figure below shows how the force on the ball from the cricketer's hand varies with time. The ball starts from rest and is thrown horizontally to another player.

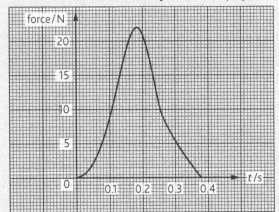

   a   Estimate the area under the graph. [2]

   b   What change in physical quantity does the area under the graph represent? [1]

   c   Calculate the horizontal velocity of the ball when it is released. [2]

   d   Calculate the maximum horizontal acceleration of the ball. [2]

   e   Sketch the force–time graph to show the force exerted by ball on the player's hand. [2]

5   a   Explain what is meant by the term 'impulse'. [1]

   b   Distinguish between mass and weight. [2]

   c   State Newton's third law of motion. [2]

   d   Use Newton's laws to explain how a rocket is able to leave the Earth's surface. [3]

6   a   Define linear momentum. [1]

   b   State the law of conservation of linear momentum. [3]

   c   An object of mass 4 kg travelling at 5 m s^{-1} strikes another object of mass 2 kg travelling at 1 m s^{-1} in the opposite direction. The objects stick together and move off with a velocity v. Calculate v. [3]

7   The north poles of two bar magnets are held together. When released, the magnets move off in opposite directions. Explain how the principle of conservation of momentum is applied to this situation. [3]

8   State two physical quantities that are conserved in an elastic collision. [2]

9   A skydiver has a mass of 80 kg. He jumps from an aircraft and free falls. He reaches a terminal velocity of 80 m s^{-1} before opening his parachute. Calculate:

   a   the weight of the skydiver [1]

   b   the air resistance F acting on the skydiver when travelling at terminal velocity [1]

   c   the magnitude of k if $F = kv^2$. [2]

   d   the acceleration of the skydiver when his velocity is 42 m s^{-1}. [3]

10 a   Explain what is meant by terminal velocity. [2]

   b   Explain why a small metal sphere falling through a viscous oil eventually reaches a terminal velocity. [3]

   c   An object has a mass of 1.9 kg. When the object falls in air, the air resistance F is given by $F = kv^2$, where v is the velocity of the object and $k = 0.028$ N s^2 m^{-2}.

   Calculate:

     i   the weight of the object [2]

     ii   the terminal velocity of the object [3]

     iii   the acceleration of the object when it is falling with a velocity of 5 m s^{-1}. [3]

11 A cuboid with dimensions 30 cm × 25 cm × 15 cm
and a mass of 3.8 kg is floating in water of density
1×10^3 kg m^{-3} so that its largest faces are horizontal.
Calculate:
   a   the upthrust on the cuboid   [3]
   b   the fraction of the cuboid that is beneath the
       surface of the water.   [2]

12 a   State two conditions for a body to be in
       equilibrium.   [2]
   b   Three co-planar force A, B and C act on a body
       that is in equilibrium.
       i   Explain how a vector triangle can be used to
           represent the forces A, B and C.   [3]
       ii Explain how the triangle illustrates that the
           forces A, B and C are in equilibrium.   [1]
   c   A toy of mass 0.75 kg hangs from two strings as
       shown below.

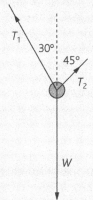

       The toy is in equilibrium. Draw a vector triangle
       to determine the magnitudes of T_1 and T_2.   [4]

13 Explain what is meant by the centre of gravity of a
   body.   [2]

14 Distinguish between the moment of a force and the
   torque of a couple.   [2]

15 A uniform plank of weight 90 N and length 2.00 m
   rests on two supports A and B. The supports A and
   B are located 0.2 m from each end of the plank. A
   construction worker of weight 800 N stands 0.45 m
   from one end. (Side closer to support B)
   a   Sketch a diagram of the plank to show the forces
       acting on it.   [2]
   b   Calculate the force acting on the plank at support
       B.   [3]
   c   Calculate the force acting on the plank at support
       A.   [2]

16 State two conditions necessary for a body to be in
   equilibrium.   [2]

17 Three forces act on an object O as shown in the
   figure below. Find the resultant of these forces and
   its direction with respect to the horizontal.

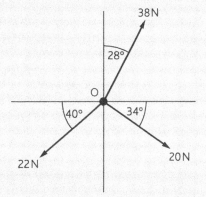

18 A car of mass 500 kg is travelling along a horizontal
   road with a constant velocity of 10 m s^{-1}. The car
   then descends a hill of length 300 m while travelling
   through a vertical distance of 20 m. A constant
   frictional force of 200 N acts on the car as it moves
   down the hill. Calculate:
   a   the initial kinetic energy of the car
   b   the total energy possessed by the car at the top
       of the hill
   c   the work done by the frictional force
   d   the velocity of the car at the bottom of the hill.

47

7 Circular motion

7.1 Motion in a circle

Learning outcomes

On completion of this section, you should be able to:

- express angular displacement in radians
- understand the concept of angular velocity
- understand the concepts of centripetal force and centripetal acceleration.

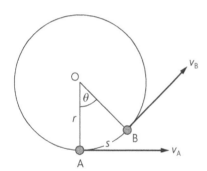

Figure 7.1.1 *Defining the radian*

Definition

The radian is defined as the angle subtended at centre of a circle by an arc equal in length to the radius of the circle.

Definition

Angular velocity ω, is defined as the rate of change of angular displacement. The SI unit is rad s^{-1}.

Equation

$v = r\omega$

v – linear velocity/m s^{-1}
r – radius/m
ω – angular velocity/rad s^{-1}

Angular velocity

Consider an object attached to a string of length r (Figure 7.1.1). The object is made to travel in a circular path at a constant speed v. The object is initially at the point A. As it travels in an anticlockwise direction an angle is swept out. When the object reaches the point B, an angle of θ is swept out. The distance travelled along the arc of the circle is s. An angle of one **radian** is defined such that $s = r$.

Therefore if the radius of the circle is r and the arc length is s, then $\theta = \dfrac{s}{r}$

When the object returns to the point A, the distance travelled will be the circumference of the circle.

Therefore, the arc length in one revolution $s = 2\pi r$

$$\therefore \qquad \theta = \frac{s}{r} = \frac{2\pi r}{r} = 2\pi$$

Therefore, one complete revolution is equivalent to 2π radians.

$$\pi \text{ radians} = 180° \qquad \textbf{1 radian} = \textbf{57.3°}$$

The angular velocity ω is the rate of change of angular displacement.

$$\omega = \frac{\theta}{t}$$

The time taken to complete on revolution is called the **period** T.

$$t = \frac{\theta}{\omega}$$

The angular displacement during one revolution is 2π.

$$\therefore \qquad T = \frac{2\pi}{\omega}$$

The linear velocity of the object at any point in the circle is given by $v = \dfrac{s}{t}$

Since the angular velocity is ω, the angle swept out in time t is ωt.

But, $\qquad \omega t = \dfrac{s}{r}$

Therefore, $\qquad v = \dfrac{s}{t} = \dfrac{\omega t r}{t} = r\omega$

Centripetal acceleration

An object moving in a straight line at a constant speed is not accelerating. However, an object moving in a circular path at a constant speed is accelerating. Consider an object travelling at a constant speed v in a circular path as shown in Figure 7.1.2. The speed of the object at any point on the circular path remains unchanged. The direction of the object is continuously changing. Since velocity is a vector quantity, the velocity at the point A, v_A is different from the velocity at the point B, v_B. The change in velocity is found using vector subtraction.

Change in velocity $\Delta v = v_B - v_A$

Vector subtraction can be thought of as a vector addition as follows:

$$\Delta v \;=\; v_B - v_A \;=\; v_B + (-v_A)$$

In order to perform the addition, the vector v_B is first drawn. The vector $-v_A$ is then drawn. The starting point for this vector is the ending point of vector v_B.

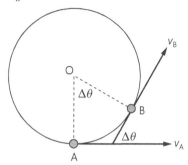
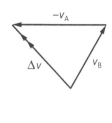

Figure 7.1.2 *Uniform circular motion*

$$\Delta v \;=\; v_B - v_A \;=\; v_B + (-v_A)$$

Acceleration is the rate of change of velocity.

$$a \;=\; \frac{\Delta v}{\Delta t}$$

If Δt is small, $\Delta\theta$ is small and $\Delta v = v\Delta\theta$.

$$a \;=\; \frac{v\Delta\theta}{\Delta t} \;=\; v\omega$$

But $v = r\omega$ $\quad \therefore\; a = (r\omega)\omega = \omega^2 r$

This acceleration is always directed towards the centre of the circle.

Centripetal force

It has already been established that the acceleration of an object travelling in a circular path with constant speed is directed towards the centre of the circle. According to Newton's second law of motion, a force is required to produce this acceleration. This force is called the **centripetal force**. This force must also act towards the centre of the circle.

According to Newton's second law $F = ma$.

But $a = \dfrac{v^2}{r}$ $\qquad \therefore\; F = \dfrac{mv^2}{r}$

Since $a = \omega^2 r$, the centripetal force can be written as $F = m\omega^2 r$.

Example

An object of mass 1.2 kg is travelling in a circular path of radius 0.8 m, with a constant speed of 0.5 m s⁻¹. Calculate:

a the angular velocity of the object
b the time taken for the object to complete one revolution
c the centripetal acceleration of the object
d the centripetal force acting on the object.

a $\omega = \dfrac{v}{r} = \dfrac{0.5}{0.8} = 0.625\,\mathrm{rad\,s^{-1}}$

b $T = \dfrac{2\pi}{\omega} = \dfrac{2\pi}{0.625} = 10.1\,\mathrm{s}$

c $a = \omega^2 r = (0.625)^2(0.8) = 0.313\,\mathrm{m\,s^{-2}}$

d $F = ma = 1.2 \times 0.313 = 0.375\,\mathrm{N}$

Equation

$$a = \omega^2 r$$

a – acceleration/m s⁻²
ω – angular velocity/rad s⁻¹
r – radius/m

Equation

$$F = \frac{mv^2}{r}$$

F – centripetal force/N
m – mass/kg
v – velocity/m s⁻¹
r – radius/m

Key points

- Angular velocity is the rate of change of angular displacement.

- An object travelling in a circular path at a constant speed is accelerating.

- The direction of the object is changing and hence its velocity is changing.

- The object is accelerating and is directed towards the centre of the circle.

- An unbalanced force called the centripetal force is required to produce a centripetal acceleration.

Learning outcomes

On completion of this section, you should be able to:

- analyse motion in a horizontal circle

- analyse motion in a vertical circle

- analyse the motion of a conical pendulum.

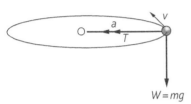

Figure 7.2.1 *An object moving in a horizontal circle*

✅ Exam tip

Make sure that you understand the diagrams and method used to calculate the tension in the string, when an object is whirled in a vertical circle.

An object attached to a string being whirled in a circular path

If a stone is attached to a string and whirled in a circular path, the tension in the string provides the centripetal force necessary to maintain the circular motion.

Horizontal circle

Consider an object of mass m, attached to a string of length r being whirled in a horizontal circle with a constant speed v as in Figure 7.2.1.

As mentioned earlier, the tension T in the string provides the centripetal force needed for the object to move in a circular path. The acceleration of the object is directed towards the centre of the circle. According to Newton's second law

$$F = ma$$

$$\therefore \quad T = \frac{mv^2}{r}$$

If the string breaks, the tension T will not be present. As a result, the centripetal force acting on the object will no longer exist. According to Newton's first law, the object will fly off in a straight line (tangent to the circle), in the direction of the instantaneous velocity at the time when the string breaks. The object does not move off in a direction radially away from the centre of the circle.

Vertical circle

Consider an object of mass m, attached to a string of length r being whirled in a vertical circle with a constant speed v. Figure 7.2.2 shows the object at different positions and how the tension in the string is determined. The tension is at a minimum at the top of the circle. The tension is at a maximum at the bottom of the circle.

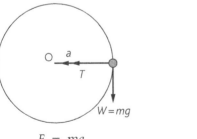

$$F = ma$$
$$\therefore \quad T = \frac{mv^2}{r}$$

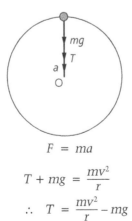

$$F = ma$$
$$T + mg = \frac{mv^2}{r}$$
$$\therefore \quad T = \frac{mv^2}{r} - mg$$

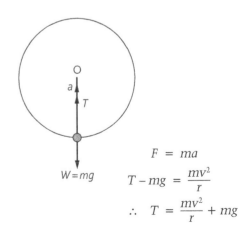

$$F = ma$$
$$T - mg = \frac{mv^2}{r}$$
$$\therefore \quad T = \frac{mv^2}{r} + mg$$

Figure 7.2.2 *An object moving in a vertical circle*

Example

An object of mass 0.80 kg is attached to a string and spun in a vertical circle of radius 0.90 m with a constant speed of 9 m s⁻¹. Calculate:

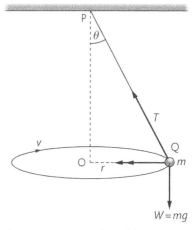

a the minimum tension in the string

b the maximum tension in the string.

a Minimum tension occurs when the mass is at the highest point in the circular path.

Minimum tension $T = \dfrac{mv^2}{r} - mg = \dfrac{0.80(9)^2}{0.90} - (0.80)(9.81)$

$= 64.2\,\text{N}$

b Maximum tension occurs when the mass is at the lowest point in the circular path.

Maximum tension $T = \dfrac{mv^2}{r} + mg = \dfrac{0.80(9)^2}{0.90} + (0.80)(9.81)$

$= 79.8\,\text{N}$

Figure 7.2.3 A conical pendulum

A conical pendulum

Consider an object of mass m attached to a string PQ of length l. The point P is fixed to a support and the mass is made to rotate in a horizontal circle of radius r shown in Figure 7.2.3. When the velocity of the object is constant, the string makes an angle of θ to the vertical.

Figure 7.2.4 shows the forces acting on the object.

Since the mass is moving in a circular path, there must be a centripetal force acting toward the centre of the circle O. The horizontal component of the tension, $T\sin\theta$ provides the centripetal force.

$$F = ma$$

$$\therefore \qquad T\sin\theta = \frac{mv^2}{r} \qquad (1)$$

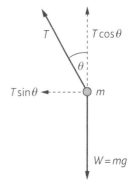

Figure 7.2.4 Analysing the forces acting on the object

The mass does not move in a vertical direction.

$$\therefore \qquad T\cos\theta = mg \qquad (2)$$

Equation (1) divided by (2)

$$\frac{T\sin\theta}{T\cos\theta} = \frac{mv^2}{r} \div mg$$

$$\tan\theta = \frac{v^2}{rg}$$

Example

A small mass of 60 g is attached to a string. One end of the string is fixed to a rigid support. The mass is made to travel in a horizontal circle of radius 0.18 m. The string makes an angle of 60° to the vertical. The mass takes 0.65 s to complete one revolution. Calculate:

a the angular velocity of the mass

b the centripetal acceleration of the mass

c the centripetal force acting on the mass

d the tension in the string.

a $T = 0.65s, \; \omega = \dfrac{2\pi}{T} = \dfrac{2\pi}{0.65} = 9.67\,\text{rad s}^{-1}$

b $a = \omega^2 r = (9.67)^2(0.18) = 16.8\,\text{m s}^{-2}$

c $F = ma = 0.06 \times 16.8 = 1.01\,\text{N}$

d $T\cos 60° = 0.06 \times 9.81$

$\qquad T = \dfrac{0.06 \times 9.81}{\cos 60°} = 1.18\,\text{N}$ (Figure 7.2.5)

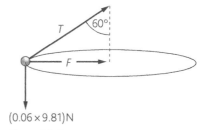

$(0.06 \times 9.81)\,\text{N}$

Figure 7.2.5

Key points

■ For a mass attached to a string the tension in the string provides the centripetal force required to keep an object moving in a circular path.

■ The tension acting on a mass being whirled in a horizontal circle with a constant speed is constant.

■ The tension acting on a mass being whirled in a vertical circle with a constant speed varies.

■ When analysing a conical pendulum it is necessary to resolve the tension into its vertical and horizontal components.

Vehicles going around a bend

In order for a car to go around a bend (arc of a circle), the friction between the tyres and the road provide the necessary centripetal force. When roads are being designed, care is taken to ensure that cars do not skid off the road while going around bends. Instead of making curved roads flat, they are banked. Figure 7.3.1 shows a vehicle on a banked road.

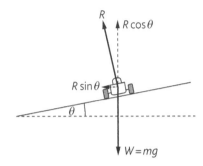

Figure 7.3.1 *A vehicle travelling along a banked road*

Using Newton's second law $F = ma$

$$\therefore \qquad R\sin\theta = \frac{mv^2}{r} \qquad (1)$$

Assuming the vehicle does not move in a vertical direction.

$$R\cos\theta = mg \qquad (2)$$

Equation (1) divided by Equation (2)

$$\frac{R\sin\theta}{R\cos\theta} = \frac{mv^2}{r} \div mg$$

$$\tan\theta = \frac{v^2}{rg}$$

An aircraft banking

An aircraft flying horizontally experiences a lift force L which balances the weight W of the aircraft. In order for the aircraft to turn, it tilts its wings at an angle θ to the vertical. As the aircraft banks, the horizontal component of the lift, $L\sin\theta$ provides the necessary centripetal force to make the aircraft turn.

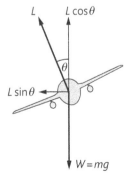

Figure 7.3.2 *An aircraft banking*

Using Newton's second law $\quad F = ma$

$\therefore \qquad L\sin\theta = \dfrac{mv^2}{r}$ $\hfill$ (3)

Assuming the aircraft does not move in a vertical direction.

$\qquad L\cos\theta = mg$ $\hfill$ (4)

Equation (3) divided by Equation (4)

$$\frac{L\sin\theta}{L\cos\theta} = \frac{mv^2}{r} \div mg$$

$$\tan\theta = \frac{v^2}{rg}$$

Example

An aircraft of mass $3.5 \times 10^4\,$kg flies with its wings tilted in order to fly in a horizontal direction of radius r. It is travelling at a constant speed of $200\,$m s^{-1}.

Calculate:

a the vertical component of L

b the lift force L

c the horizontal component of L

d the acceleration of the aircraft towards the centre of the circle

e the value of r.

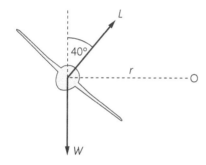

Figure 7.3.3

a Vertical component L = weight of aircraft

$\qquad\qquad W = mg = 3.5 \times 10^4 \times 9.81 = 3.43 \times 10^5\,$N

b $L\cos 40° = 3.43 \times 10^5$

$\qquad L = \dfrac{3.43 \times 10^5}{\cos 40°} = 4.48 \times 10^5\,$N

c Horizontal component of $L = L\cos(90-40)°$

$\qquad\qquad\qquad\qquad = 4.48 \times 10^5 \times \cos 50° = 2.88 \times 10^5\,$N

d $a = \dfrac{F}{m} = \dfrac{2.88 \times 10^5}{3.5 \times 10^4} = 8.23\,$m s^{-2}

e $a = \dfrac{v^2}{r}$

$\qquad r = \dfrac{v^2}{a} = \dfrac{(200)^2}{8.23} = 4.86 \times 10^3\,$m $\qquad$ (Figure 7.3.3)

Key points

■ The friction between the tyres and the road provides the centripetal force required to keep a car moving in a circular path.

■ The horizontal component of an aircraft's lift, while banking, provides the necessary centripetal force required for it to travel along an arc.

8.1 Gravitational field

Learning outcomes

On completion of this section, you should be able to:

- understand the concept of a gravitational field
- define gravitational field strength
- state Newton's law of gravitation
- describe an experiment to determine the acceleration due to gravity.

Definition

A gravitational field is a region around a body where a mass experiences a force when placed in the field.

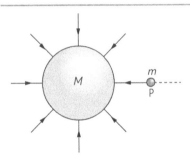

Figure 8.1.1 Diagram showing the gravitational field around the Earth

Definition

The direction of a gravitational field is the direction of the force on a test mass placed in the field.

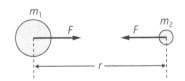

Figure 8.1.2 Newton's law of gravitation

Did you know?

If there is no negative sign, it should be stated that the force is attractive.

Gravitational field and field lines

A gravitational field exists around bodies that have mass. If an object is placed inside this field, it experiences a force. This force is attractive in nature.

Figure 8.1.1 illustrates the gravitational field around spherical object such as the Earth. If an object P of mass m, is placed inside the gravitational field of the Earth, it experiences a force. The direction of this force is towards the centre of the Earth.

The gravitational field around the Earth is represented by using field lines. The spacing of the field lines gives an idea of the strength of the field. The closer the field lines are, the stronger the field. The more spaced out the field lines are, the weaker the field.

The direction of the gravitational field is the direction of the force on a test mass placed in the field.

Newton's law of gravitation

All bodies that have mass exert a force on each other. The Earth exerts an attractive force on the Moon. The Moon exerts an equal and opposite force on the Earth according to Newton's third law. This force is what keeps the Moon in orbit around the Earth.

Consider two bodies having masses of m_1 and m_2 respectively and separated by a distance r. Newton stated that there exists a force of attraction between these two bodies. The magnitude of the force is directly proportional to the product of the masses of the bodies. It is also inversely proportional to the square of the distance between the two bodies (Figure 8.1.2).

Newton's law can be expressed as follows.

Definition

Newton's law of gravitation states that the force of attraction between any two bodies is directly proportional to the product of their masses and inversely proportional to the square of the distance between them.

Equation

$$F = -\frac{Gm_1m_2}{r^2}$$

G – gravitational constant ($6.67 \times 10^{-11}\,\mathrm{N\,m^2\,kg^{-2}}$)
m_1 – mass of one body/kg
m_2 – mass of other body/kg
r – distance between the centres of mass of the two bodies/m

The constant of proportionality G is called the gravitational constant. It has been experimentally determined as $6.67 \times 10^{-11}\,\mathrm{N\,m^2\,kg^{-2}}$. The minus sign in the equation indicates that the force is attractive. It is important to remember that Newton's third law applies. This means that if one body A, is exerting a force F on another body B, then body B will exert an equal and opposite force on body A.

Gravitational field strength

The gravitational field strength is the force acting per unit mass. This means that the gravitational field strength is the force exerted on a 1 kg mass placed in the field. On the Earth's surface the gravitational field strength is $g = 9.81\,\text{N}\,\text{kg}^{-1}$. Gravitational field strength is a vector quantity.

In Figure 8.1.1, the force exerted on object P is $F = -\dfrac{GMm}{r^2}$, where r is the distance between the centre of mass of M and m.

The gravitational field strength at P due to the mass M is g.

The force exerted on P in terms of the gravitational field strength is $F = mg$.

$$mg = -\frac{GMm}{r^2}$$

$$\therefore \qquad g = -\frac{GM}{r^2}$$

From this equation, it can be seen that the gravitational field strength at the point P is dependent on the mass of the object creating the field and distance from its centre of mass.

The acceleration due to gravity

When an object is released it falls to the ground. The force of gravity acts on the object. The gravitational force produces an acceleration which is equal to $9.81\,\text{m}\,\text{s}^{-2}$. This magnitude can be determined experimentally as follows. An iron bearing is made to fall through a known distance h and the time t taken is recorded.

$$s = ut + \frac{1}{2}at^2$$

$$s = h, \quad u = 0 \quad a = g$$

$$h = (0)t + \frac{1}{2}gt^2 = \frac{1}{2}gt^2$$

$$\therefore \qquad g = \frac{2h}{t^2}$$

Two light gates are set up as shown in Figure 8.1.3 and the vertical distance between them is measured. The metal ball is held in place using an electromagnet. When the electromagnet is switched off, the metal ball begins falling. As the ball passes through the first light gate, the timer starts. When the ball passes through the second light gate, the timer stops. The height h and the measured time t are used to calculate the acceleration due to gravity g.

Key points

- A gravitational field is the region around a body where a mass experiences a force.

- The direction of a gravitational field is the direction of the force on a test mass placed in the field.

- Newton's law of gravitation states the force of attraction between two bodies is proportional to the product of their masses and inversely proportional to the square of the distance between them.

- Gravitational field strength is the force per unit mass.

- The acceleration due to gravity is $9.81\,\text{m}\,\text{s}^{-2}$ and can be determined by measuring the time taken for a mass to travel through a known vertical distance.

Definition

The gravitational field strength is the force acting per unit mass.

Equation

$$g = \frac{F}{m}$$

g – gravitational field strength/N kg^{-1}

F – force/N

m – mass/kg

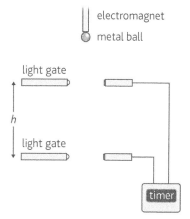

Figure 8.1.3 *Measuring the acceleration due to gravity (free fall method)*

Did you know

g varies around the Earth. The Earth is not a perfect sphere. It is squashed at the poles.

8.2 Gravitational potential and satellites

Learning outcomes

On completion of this section, you should be able to:

- define gravitational potential
- understand the term *equipotential*
- discuss the motion of geostationary satellites
- state the applications of geostationary satellites.

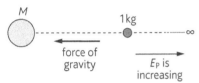

Figure 8.2.1 *Moving a 1 kg mass to a point far away from the Earth*

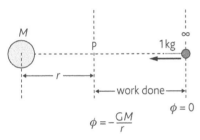

Figure 8.2.2 *Defining gravitational potential*

Definition

The gravitational potential ϕ, at a point is the work done in moving unit mass from infinity to that point.

Equation

$$\phi = -\frac{GM}{r}$$

ϕ – gravitational potential/J kg⁻¹
G – gravitational constant
$\quad$ $(6.67 \times 10^{-11} \, \text{N m}^2 \, \text{kg}^{-2})$
M – mass/kg
r – distance from the centre of mass/m

Gravitational potential

When an object is present in a gravitational field, it possesses gravitational potential energy. If an object of mass m is moved from one floor of a building to a higher one, while travelling through a vertical distance of h, the mass gains gravitational potential energy. Work is done against the force of gravity. The work done is equal to the gain in gravitational potential energy. The gain in gravitational potential energy is given by $\Delta E_\text{p} = mg\Delta h$.

It was assumed that the gravitational potential g remains constant as the mass is being moved vertically upwards. For distances close to the Earth's surface, this can be assumed to be true. However, as we move further away from the Earth's surface, the gravitational field strength no longer remains constant and the equation for gravitational potential energy no longer applies. Recall that the gravitational field strength due to a mass M varies with distance r as follows:

$$g = -\frac{GM}{r^2}$$

Consider a 1 kg mass moving away from the Earth's surface to some point where the gravitational field strength due to the Earth is negligible. Assume that this point is infinity. As the mass moves away from the Earth, work is done against the force of gravity. The mass therefore gains gravitational potential energy. At infinity, the mass would have its maximum gravitational potential energy.

Suppose a 1 kg mass is moved from infinity to some point P as shown in Figure 8.2.1. The movement of the mass and the force of gravity act in the same direction. Therefore, negative work is being done on the mass. The **gravitational potential** at the point P is defined as the work done in moving unit mass (1 kg) from infinity to that point. Gravitational potential is a scalar quantity. The gravitational potential at infinity is defined as being equal to zero. We have already indicated that at infinity, the gravitational potential energy is at a maximum. Therefore, at any point closer to the Earth, the gravitational potential energy will be less than zero. Therefore, gravitational potentials have negative values (Figure 8.2.2).

The gravitational potential ϕ, at a distance r from a point mass M is given by: $\phi = -\frac{GM}{r}$

Figure 8.2.3 shows the variation of gravitational potential with distance from the Earth.

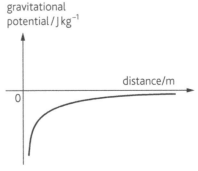

Figure 8.2.3 *Variation of gravitational potential with distance from the centre of the Earth*

Field lines and equipotentials

A gravitational field is represented using field lines. The direction of field at a point is the direction of the force acting on a point mass placed at that point. There are points within a gravitational field where the gravitational potentials are the same. A line drawn through points having the same gravitational potential is called an **equipotential** line (Figures 8.2.4 and 8.2.5).

Geostationary satellites

Satellites orbit the Earth in distinct paths. There are particular satellites called **geostationary satellites**. They orbit the Earth above the equator. They have a period of 24 hours and orbit at a distance of 3.6×10^4 km above the Earth's surface. Since the satellite has a period of 24 hours (1 day), it appears to be stationary above the same point on the equator all the time. These satellites orbits from west to east. Geostationary satellites have many uses.

Geostationary satellites are used in:

- weather monitoring
- television transmission.
- telephone communication

Global positioning satellites (GPS) are not geostationary. They have a period of 12 hours. They orbit at a height of approximately 2.02×10^4 km.

Global positioning satellites are used:

- for time synchronisation
- in cellular telephony
- to determine precise location on the Earth.
- to track vehicles
- to guide missiles

Example

A satellite of mass 2500 kg is placed in a geostationary orbit at a distance of 4.23×10^7 m from the centre of the Earth.

Calculate:

i the angular velocity of the satellite
ii the speed of the satellite in its orbit
iii the acceleration of the satellite
iv the force exerted by the Earth on the satellite
v the mass of the Earth.

i The period of a geostationary satellite = 24 hours
Angular velocity $\omega = \dfrac{2\pi}{T} = \dfrac{2\pi}{24 \times 3600} = 7.27 \times 10^{-5}\,\text{rad s}^{-1}$

ii Speed of satellite $v = r\omega$
$= 4.23 \times 10^7 \times 7.27 \times 10^5 = 3.08 \times 10^3\,\text{m s}^{-1}$

iii Acceleration of satellite $a = -\omega^2 r$
$= (7.27 \times 10^{-5})^2 \times 4.23 \times 10^7 = 0.224\,\text{m s}^{-2}$

iv Force exerted by the Earth on the satellite $F = ma$
$= 2500 \times 0.224$
$= 560\,\text{N}$

v Using $F = -\dfrac{GMm}{r^2}$
Mass of Earth $= \dfrac{Fr^2}{Gm} = \dfrac{560 \times (4.23 \times 10^7)^2}{6.67 \times 10^{-11} \times 2500} = 6.00 \times 10^{24}\,\text{kg}$

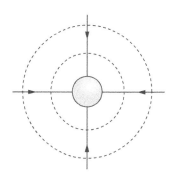

equipotential lines ---------
field lines ⟶

Figure 8.2.4 Diagram showing field lines around a spherical body

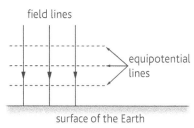

field lines

equipotential lines

surface of the Earth

Figure 8.2.5 Diagram showing field lines close to the surface of the Earth

Key points

- The gravitational potential at a point is the work done in moving unit mass from infinity to that point.

- An equipotential line is a line drawn through points having the same gravitational potential.

- A geostationary satellite has a period of 24 hours and appears to be at the same point above the Earth all the time.

Revision questions 3

Answers to questions that require calculation can be found on the accompanying CD.

1   Explain what is meant by:
   a   energy   [1]
   b   the principal of conservation of energy [3]
   c   work.   [2]

2   a   Explain what is meant by the terms 'work' and 'power'. [2]
   b   At an amusement park in Trinidad, a ride consists of a carriage being pulled up a ramp by a steel cable. The carriage and the passengers have a combined mass of 480 kg. Initially, the carriage is being pulled such that it is travelling at $8 \, m \, s^{-1}$ towards the ramp which is inclined at 30° to the horizontal. The carriage is brought to rest after travelling for some time up the slope. During the process, the carriage travels a vertical distance of 3.0 m.
      Calculate:
      i   the initial kinetic energy of the carriage and the passengers [3]
      ii the gain in gravitational potential energy of the carriage and the passengers [2]
      iii the work done against the resistive force F acting on the carriage as it moves up the ramp. [1]
      iv the magnitude of F. [2]

3   Explain what is meant by the concept of work. Use your answer to derive an expression for the increase in gravitational potential energy when an object of mass m is raised vertically through a distance h near the Earth's surface. [4]

4   A force F is acting on a body that is moving with a velocity v in the direction of the force. Derive an expression relating power P dissipated by the force to F and v. [2]

5   a   Define the radian. [1]
   b   Convert the following to radians
      i   30°   ii   140° [2]
   c   Convert the following to degrees
      i   3.14 radians   ii   1.57 radians [2]

6   a   Explain what is meant by angular velocity. [2]
   b   Describe qualitatively how it is that a body which is travelling in a circular path with uniform speed has acceleration. 3]
   c   Derive the equation for circular motion $a = \omega^2 r$, where a is the centripetal acceleration, ω is the angular velocity and r is the radius of the circle. [5]

7   An object of mass 0.85 kg is travelling in a horizontal circular path of radius 0.5 m, with a constant speed of $1.2 \, m \, s^{-1}$. Calculate:
   a   the angular velocity of the object [2]
   b   the time taken for the object to complete one revolution [2]
   c   the centripetal acceleration of the object [2]
   d   the centripetal force acting of the object. [2]

8   Calculate the force required to keep a mass of 1.2 kg revolving in a horizontal circle of radius 0.6 m with a period of 0.8 seconds. [3]

9   An object of mass 200 g is attached to a string and spun in a vertical circle of radius 0.50 m with a constant speed of $6 \, m \, s^{-1}$. Calculate:
   a   the minimum tension in the string [3]
   b   the maximum tension in the string. [3]

10 A small mass of 80 g is attached to a string. One end of the string is fixed to a rigid support. The mass is made to travel in a horizontal circle of radius 0.60 m. The string makes an angle of 40° to the vertical. The mass takes 0.15 s to complete one revolution. Calculate:
   a   the angular velocity of the mass [2]
   b   the centripetal acceleration of the mass [2]
   c   the centripetal force acting on the mass [2]
   d   the tension in the string. [2]

11 a   A mass of 0.50 kg is attached to a string and whirled in a horizontal circle of radius 1.10 m. The string will break when the tension exceeds 50 N. Calculate the maximum frequency of rotation. [5]
   b   Describe the motion of the mass if the string breaks. [3]

12 a   State Newton's law of gravitation. [2]
   b   The Earth can be considered to be a uniform sphere of radius R. R is assumed to be 6.4×10^6 m. A geostationary satellite is orbiting the Earth.
      i   Explain what is meant by a geostationary orbit. [3]
      ii Show that the radius of a geostationary orbit is given by the expression

$$r = \sqrt[3]{\frac{gR^2}{\omega^2}},$$

      where g is the acceleration due to gravity at the Earth's surface and ω is the angular

velocity of the satellite about the centre of the Earth. [3]

Determine the radius of a geostationary orbit. [3]

13 **a** Explain what is meant by:

   i gravitational potential [1]

   ii equipotential. [1]

  b Explain why gravitational potentials are always negative. [3]

14 A communications satellite is located at a height of 385 km above the Earth. The mass of the satellite is 4.2×10^3 kg. The radius of the Earth is assumed to be 6370 km. The Earth is assumed to be a point mass of 6.0×10^{24} kg. Calculate:

  a the force acting on the satellite [2]

  b the centripetal acceleration [2]

  c the speed of the satellite. [2]

15 A global positioning system (GPS) uses a number of satellites that orbit the Earth in circular orbits at a distance of 2.22×10^4 km above the Earth's surface. Calculate the angular speed of one such satellite.

Mass of Earth $= 5.99 \times 10^{24}$ kg

Radius of Earth $= 6.38 \times 10^3$ km [3]

16 The Earth may be assumed to be a sphere of mass 6.0×10^{24} kg. The Moon may also be considered a sphere of mass 7.35×10^{22} kg. The distance from the centre of the Earth to the centre of the Moon is 3.84×10^8 m. Assume that the Moon travels at a constant speed in a circular orbit around the Earth.

  i Calculate the gravitational force exerted by the Earth on the Moon.

  ii Calculate the acceleration of the Moon.

  iii Sketch a diagram showing the direction of this acceleration.

  iv Explain why this acceleration does not increase the speed of the Moon.

  v Determine the gravitational field strength of the Earth at the Moon.

Answers to the multiple-choice questions and to selected structured questions can be found on the accompanying CD.

Multiple-choice questions

1 Which of the following pairs of units are SI base units?
   a ampere, degree celsius
   b coulomb, kelvin
   c kilogram, kelvin
   d metre, degree Celsius

2 A student wishes to measure the density of material X. He has cube of material X. He measures the mass and the average length of one side of the cube.

   Mass of cube = $16.5 \pm 0.5\,g$

   Length of one side = $4.2 \pm 0.1\,cm$

   What is the percentage error when the student determines the density of material X?
   a 3%   b 2%   c 9%   d 5%

3 The SI unit for specific heat capacity in terms of base units is:
   a $kg\,m^2\,s^{-2}\,K^{-1}$ b $m^2\,s^{-2}\,K^{-1}$
   c $m^2\,s^2\,K^{-1}$ d $m^{-2}\,s^{-2}\,K^{-1}$
   (Refer to 15.1.)

4 What is the number of atoms present in 0.090 kg of carbon-12?
   a 4.5×10^{22} b 4.5×10^{21}
   c 4.5×10^{23} d 4.5×10^{24}

5 What is the ratio $\dfrac{1\,\mu g}{1\,kg}$?
   a 10^{-2} b 10^{-12} c 10^{-9} d 10^{-3}

6 A small object is projected horizontally from a wall of height 7 m with a speed of $30\,m\,s^{-1}$. What is the velocity of the object just before striking the ground?
   a $32.2\,m\,s^{-1}$ b $30.0\,m\,s^{-1}$
   c $11.8\,m\,s^{-1}$ d $34.2\,m\,s^{-1}$

7 A uniform plank of length l is supported by two straps as shown below. What is the ratio $\dfrac{T_1}{T_2}$?

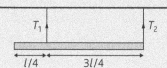

   a ½   b 2   c ¼   d 4

8 Which of the following is not true about inelastic collisions?
   a Momentum is conserved
   b Total energy is conserved
   c Kinetic energy is conserved
   d Kinetic energy is not conserved

9 A compact disc (CD) is placed inside a CD player and starts to rotate. A point Y is at the outer edge of the disc. A point X on the disc is ½ the distance as the point Y is from the centre of the axis of rotation. The linear velocity of Y is a. The linear velocity of X is b. At any point in time, the ratio a/b is
   a ½   b 2   c 4   d ¼

10 A satellite has a weight W before launch. It is then placed in orbit at a height $h = 5R$ above the Earth's surface. R is the radius of the Earth. What is the gravitational force acting on the satellite when it is orbiting the Earth?
   a $\dfrac{W}{5}$ b $\dfrac{W}{6}$ c $\dfrac{W}{25}$ d $\dfrac{W}{36}$

Structured questions

11 a Distinguish between precision and accuracy.   [2]
   b Explain what is meant by a systematic and a random error and give an example of each type of error.   [4]
   c A student is given five marbles and asked to determine the density of the material used to make the marbles. The student decides to line the marbles up in a straight line against the edge of a metre rule. She notes the beginning and ending point along the ruler. She then uses a balance to measure the mass of the five marbles. Her results are as follows:

   $X = 12.5 \pm 0.1\,cm$

   $Y = 20.0 \pm 0.1\,cm$

   Mass of five marbles = $20.5 \pm 0.5\,g$

   Calculate:
   i the diameter of one marble, including the absolute uncertainty   [2]
   ii the mass of one marble, including the absolute uncertainty   [2]
   iii the density of the material used to make the marbles, including the uncertainty.   [3]

12 A cyclist is training in a hilly region in Jamaica. The total mass of the cyclist and his bicycle is 85 kg. Initially he is travelling at a constant speed of $12\,m\,s^{-1}$ on a level dirt road. He then travels down a slope on to another level road while travelling through a vertical distance of 5.0 m.

Calculate:

a the kinetic energy of the cyclist and his bicycle on the level road [2]

b the loss in potential energy while travelling through a vertical distance of 5.0 m [2]

c the speed of the cyclist and his bicycle at the bottom of the slope. [3]

Given that the cyclist was providing a power of 320 W when he was travelling at $12\,m\,s^{-1}$, calculate the total resistive force acting on the cyclist. [2]

13 a Explain what is meant by the terms 'work' and 'energy'. [2]

b A car of mass 900 kg is travelling at a constant speed of $18\,m\,s^{-1}$ down a sloped road. The angle of the road to the horizontal is 12°. The driver notices another vehicle in front of her. She applies the brakes to bring the car to a complete stop. A constant force of 3000 N opposes the motion of the car.

i Sketch a diagram to show forces acting on the car when it is at rest on the slope. [2]

ii Calculate the component of the weight of the car down the slope. [2]

iii Calculate the normal reaction acting on the car. [2]

iv Calculate the deceleration of the car when the brakes are applied. [2]

v Calculate the distance travelled by the car from the point where the brakes are applied to the point where the car stops. [2]

vi Calculate the loss of kinetic energy of the car. [2]

vii Calculate the work done by the 3000 N force. [1]

14 a Distinguish between scalar and vector quantities. [2]

b Give an example of each. [2]

c A cannon in Tobago is positioned such that it lies horizontally on the edge of a cliff 20 m high. A cannon ball is fired horizontally with a velocity of $45\,m\,s^{-1}$.

Calculate:

i the time taken for the cannon ball to hit the surface of the sea [2]

ii the horizontal distance travelled by the cannon ball [2]

iii the velocity of the cannon ball just before hitting the sea. [3]

15 a State two conditions that must be satisfied for a body to be in equilibrium. [2]

b Three forces P, Q and R act on an object O. The object O is in equilibrium. Explain using a sketch how a vector triangle is drawn to represent these forces. [3]

c How does the triangle show that the object O is in equilibrium? [1]

16 a State Newton's law of gravitation. [2]

b The mass and radius of the Earth are assumed to be 5.98×10^{24} kg and 6.40×10^{6} m respectively. Determine a value for the gravitational field strength g at the Earth's surface. [3]

c A geostationary satellite is at a distance of 4.23×10^{7} m from the centre of the Earth and is orbiting above the equator.

i Explain what is meant by a geostationary satellite. [1]

ii Calculate the gravitational field strength at the point where the satellite is located. [1]

iii Calculate the speed of the satellite. [3]

iv Calculate the acceleration of the satellite. [2]

17 a Explain what is meant by linear momentum. [2]

b State the law of conservation of linear momentum. [3]

c Distinguish between an elastic collision and an inelastic collision. [2]

18 a Explain how is it that an object travelling in a circular path with uniform speed has acceleration. State the direction of the force producing this acceleration. [4]

b Derive the equation for circular motion $a = \omega^2 r$, where a is the centripetal acceleration, ω is the angular velocity and r is the radius of the circle. [4]

c A mass of 0.85 kg is attached to a string and rotated in a vertical circle of radius 1.50 m. The minimum tension in the string is 2.5 N.

i Determine speed of rotation. [3]

ii Determine the maximum tension in the string. [3]

9.1 Free oscillations

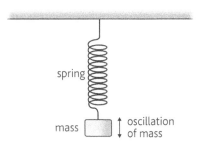

Figure 9.1.3 *A mass oscillating on a spring*

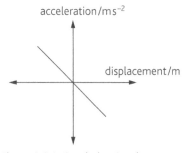

Figure 9.1.4 *Graph showing the relationship between acceleration and displacement*

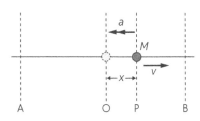

Figure 9.1.5 *Simple harmonic motion*

Examples of free oscillations

Examples of free oscillations are shown in Figures 9.1.1–3.

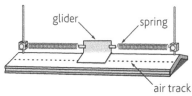

Figure 9.1.1 *A glider oscillating back and forth on an air track using two springs*

Figure 9.1.2 *A small marble oscillating in a dish*

Simple harmonic motion (SHM)

Simple harmonic motion is a periodic motion in which the acceleration of a mass is proportional to the displacement from a fixed point and directed towards the fixed point.

The following equation is used to define SHM. Note that the negative sign (–) indicates that the direction of the acceleration is always opposite to the direction of the displacement.

Definition	*Equation*
SHM is a periodic motion in which:	$a = -\omega^2 x$
1 the acceleration is proportional to the displacement from a fixed point and	a – acceleration/m s^{-2}
	ω – angular frequency/rad s^{-1}
2 directed towards the fixed point.	x – displacement/m

Figure 9.1.4 shows the relationship between acceleration and displacement for an object undergoing SHM.

Consider a particle M initially starting at the point O (equilibrium or fixed position). It begins to oscillate with SHM about the point O. The points A and B represent the maximum displacement from the equilibrium position O. A snapshot of its motion is illustrated in Figure 9.1.5. At point P the velocity and acceleration of M are v and a respectively. The displacement of M from the equilibrium position is x. It can be seen that v and a are in opposite directions.

The conditions necessary for SHM are as follows:

- A mass that oscillates.

- A fixed point at which the mass is in equilibrium.

- A restoring force which returns the mass to its equilibrium position if it is displaced.

Displacement, velocity and acceleration

Consider the motion of a simple pendulum. When the bob is displaced to one side gently and then released, gravity pulls on it. This force causes it

to return to its equilibrium position. However, the bob passes this point and causes the process to be repeated again. So the bob oscillates to the left and right of the equilibrium position.

SHM can be illustrated graphically (Figure 9.1.6). The pendulum bob is initially at O at time $t = 0$. The pendulum is displaced to the right until it reaches the point A. One oscillation is the motion of the pendulum bob as follows O $\longrightarrow$ A $\longrightarrow$ O $\longrightarrow$ B $\longrightarrow$ O. The time taken to complete one oscillation is T. This time is called the **period** of oscillation.

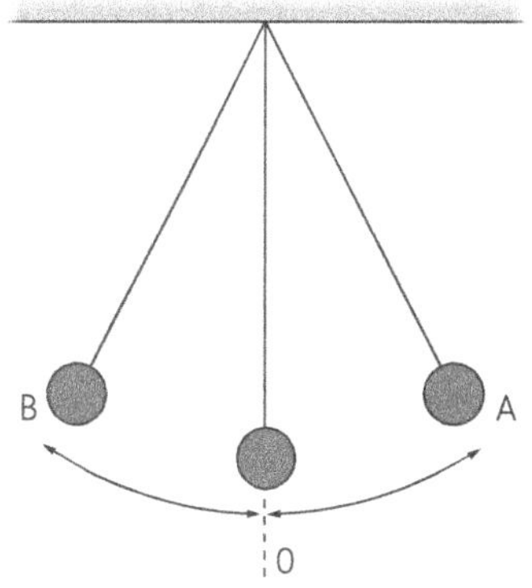

Figure 9.1.7 (Graph I) illustrates a displacement–time graph for the motion of a simple pendulum. The graph is sinusoidal in shape.

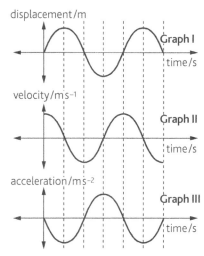

Figure 9.1.6 *The motion of a simple pendulum*

If the amplitude of the displacement (maximum distance from the equilibrium position) is A, then the displacement as a function of time is $x = A \sin \omega t$.

Equation

$x = A \sin \omega t$

x – displacement/m
A – amplitude/m
ω – angular frequency/rad s^{-1}
t – time/s

To determine the velocity at any point on a displacement–time graph the gradient of the tangent at that point must be determined. Therefore, at time $t = 0$, the gradient is a maximum and is positive. At time $t = T/4$, the gradient is zero. At time $t = T/2$, the gradient is a maximum and negative. At time $t = 3T/4$, the gradient is zero. At time $t = T$, the gradient is a maximum and positive.

The velocity–time graph (Figure 9.1.7 Graph II) can be expressed mathematically in the equation opposite.

Equation

$v = v_0 \cos \omega t$

v – velocity/m s^{-1}
v_0 – maximum velocity/m s^{-1}
ω – angular frequency/rad s^{-1}
t – time/s

Similarly, the velocity–time graph can be used to obtain the acceleration–time graph for the motion of the pendulum. The acceleration–time graph is found by finding the gradient at various points on the velocity–time graph (Figure 9.1.7 Graph III).

Figure 9.1.7 *Graphs showing the variation of displacement, velocity and acceleration with time*

The equation gives the mathematical expression for the velocity of an object undergoing SHM related to displacement (Figure 9.1.8).

Equation

$v = \pm \omega \sqrt{(x_0^2 - x^2)}$

v – velocity/m s^{-1}
ω – angular frequency/rad s^{-1}
x_0 – maximum displacement/m
x – displacement/m

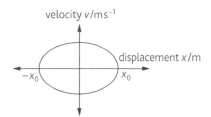

Figure 9.1.8 *Graph of velocity against displacement*

Key points

- Free oscillations include the motion of a simple pendulum and a mass attached to a spring.

- SHM is a periodic motion in which the acceleration is proportional to the displacement from a fixed point and directed towards the fixed point.

- The conditions for SHM are an oscillating mass and a restoring force.

- The displacement–time graph for an object undergoing SHM is sinusoidal.

- The velocity–time graphs and the acceleration–time graphs for an object undergoing SHM are also sinusoidal.

Definitions

Displacement is the distance moved in a stated direction from the equilibrium position.

Amplitude is the maximum displacement from the equilibrium position.

Definitions

The period is the time taken for one oscillation.

The frequency is the number of oscillations per unit time.

Equation

$$f = \frac{1}{T}$$

f – frequency/Hz
T – period/s

Displacement and amplitude

Displacement is the distance moved in a stated direction from the equilibrium position. Figure 9.2.1 shows part of a displacement–time graph for a simple pendulum. The dashed line represents the equilibrium position. The point P represents the position of the pendulum at a particular instant in time. The distance x represents the displacement. The maximum displacement from the equilibrium is called the amplitude. The SI unit for displacement and amplitude is the metre.

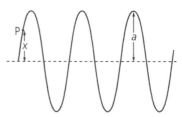

Figure 9.2.1 *Diagram illustrating displacement and amplitude*

Period, frequency and angular frequency

Suppose a simple pendulum is displaced and left to oscillate. The period T is the time taken for one oscillation. The SI unit is the second (s). Figure 9.2.2 illustrates a displacement–time graph and shows how the period T is determined.

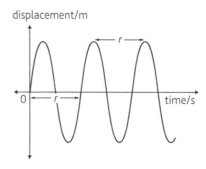

Figure 9.2.2 *Determining T from a displacement–time graph*

Frequency f is the number of oscillations per unit time. The SI unit is the hertz (Hz). If one oscillation is produced in a time T, then it follows that $1/T$ oscillations are produced in 1 second.

Angular frequency (angular velocity) ω is defined as the rate of change of angular displacement (see 7.1, *Motion in a circle*). The SI unit is rad s^{-1}.

Angular frequency and frequency are related by the following equations.

Equation

$$\omega = 2\pi f$$

ω – angular frequency/rad s^{-1}
f – frequency/Hz

Equation

$$\omega = \frac{2\pi}{T}$$

ω – angular frequency/rad s^{-1}
T – period/s

Linking simple harmonic motion and circular motion

Simple harmonic motion and circular motion are closely related. Consider a vertical peg attached to a disc of radius r, attached to a turntable. The disc rotates at a constant angular velocity ω. A horizontal beam of parallel light produces a shadow of the peg on a screen. At time $t = 0$, the shadow is at O. At time t, the disc rotates through an angle θ. The shadow moves from O to P.

$$\theta = \omega t$$

Distance $\quad OP = r \sin(\omega t)$

The shadow of the peg moves back and forth between the points A and B. The amplitude of this movement is r. The shadow moves with simple harmonic motion (Figures 9.2.3 and 9.2.4).

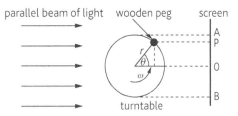

Figure 9.2.3 *Experiment showing the relationship between SHM and circular motion (side view)*

Figure 9.2.4 *Experiment to show the relationship between SHM and circular motion (top view)*

Example

The graph in Figure 9.2.5 shows how the acceleration of an object undergoing simple harmonic motion varies with time.

Determine:

a the period of oscillation

b the frequency

c the angular frequency, ω

d the amplitude x_0 of the oscillation.

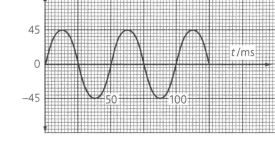

Figure 9.2.5

a Period $T = 50\,\text{ms}$

b frequency $f = \dfrac{1}{T} = \dfrac{1}{50 \times 10^{-3}} = 20\,\text{Hz}$

c angular frequency $\omega = 2\pi f = 2\pi(20) = 126\,\text{rad s}^{-1}$

d $a = -\omega^2 x$

Maximum acceleration occurs at maximum displacement (amplitude)

$x_0 = -\dfrac{a}{\omega^2} = -\dfrac{45}{(126)^2} = -2.83 \times 10^{-3}\,\text{m}$

Key points

- Displacement is the distance moved in a stated direction from the equilibrium position.
- Amplitude is the maximum displacement from the equilibrium position.
- The period is the time taken for one oscillation.
- The frequency is the number of oscillations per unit time.
- Circular motion and simple harmonic motion are closely related.

Learning outcomes

On completion of this section, you should be able to:

- derive the equation for the period of a simple pendulum

- derive the equation for the period of a mass on a spring

- describe the interchange between kinetic and potential energy during simple harmonic motion.

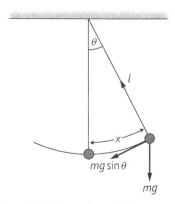

Figure 9.3.1 *A simple pendulum*

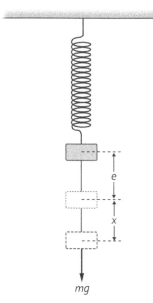

Figure 9.3.2 *A mass-spring system*

A simple pendulum

The mass m is displaced to the right through an arc x (Figure 9.3.1).

$$\text{Restoring force} = mg\sin\theta$$

For small angles (in radians) $\sin\theta \approx \theta \approx \dfrac{x}{l}$

$$\text{Therefore, the restoring force} = mg\left(\frac{x}{l}\right)$$

Using Newton's second law $\quad F = ma$

$$\text{Acceleration towards O} \quad = \frac{F}{m} = \frac{mgx}{l} \div m = \frac{gx}{l}$$

Acceleration in the direction of x, $a = -\dfrac{gx}{l}$

Comparing with the equation for simple harmonic motion $a = -\omega^2 x$

$$\frac{gx}{l} = \omega^2 x$$

$$\text{But} \qquad \omega = \frac{2\pi}{T}$$

$$\therefore \qquad \frac{g}{l} = \left|\frac{2\pi}{T}\right|^2 = \frac{4\pi^2}{T^2}$$

$$T^2 = \frac{4\pi^2 l}{g}$$

$$T = 2\pi\sqrt{\frac{l}{g}}$$

A mass attached to a spring

Consider a spring with mass m attached to it. Assume that the spring obeys Hooke's law $F = ke$, where k is the spring constant, F is the force applied and e is the extension produced (Figure 9.3.2).

$$\therefore \qquad mg = ke$$

When the mass is pulled downwards a distance x and released, it makes small oscillations in a vertical plane. The tension T in the spring is given by $k(e + x)$. The resultant downward force is

$$ke - k(e + x) = ke - ke - kx = -kx.$$

Using Newton's second law $\quad F = ma$

$$-kx = ma$$

$$\therefore \qquad a = -\frac{kx}{m} \text{ (downwards) in the direction of } x$$

Comparing with the equation for simple harmonic motion $a = -\omega^2 x$

$$\frac{k}{m} = \omega^2$$

$$\text{But } \omega = \frac{2\pi}{T}$$

$$\frac{k}{m} = \left|\frac{2\pi}{T}\right|^2 = \frac{4\pi^2}{T^2}$$

$$T^2 = \frac{4\pi^2 m}{k}$$

$$T = 2\pi\sqrt{\frac{m}{k}}$$

Rewriting in terms of e and g: $\quad T = 2\pi\sqrt{\frac{e}{g}} \qquad$ since $mg = ke$

The period of oscillation for two similar springs in series is $T = 2\pi\sqrt{\frac{2e}{g}}$

The period of oscillation for two similar springs in parallel is $T = 2\pi\sqrt{\frac{e/2}{g}}$

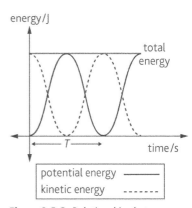

Figure 9.3.3 *Relationships between energy and time*

Energy in simple harmonic motion

When a system is oscillating with simple harmonic motion, there is an interchange between the potential and kinetic energy of the system. The total energy of the system remains constant, provided that the system is not damped. Consider the motion of a simple pendulum. When the bob is displaced to one side and released, the system oscillates with simple harmonic motion. When the bob is at its maximum displacement, its potential energy is at a maximum. The kinetic energy at this point is zero (velocity $= 0$). When the bob passes the equilibrium position, its velocity is at a maximum and therefore its kinetic energy is also at a maximum. At the equilibrium position, its potential energy is at a minimum. Figure 9.3.3 shows the variation of kinetic energy, potential energy and total energy with time for the motion of a simple pendulum.

The energy of a system oscillating with simple harmonic motion can be represented graphically as a function of displacement. At maximum displacement, the potential energy is at a maximum and the kinetic energy is at a minimum. At zero displacement the kinetic energy is at a maximum and the potential energy is at a minimum (Figure 9.3.4).

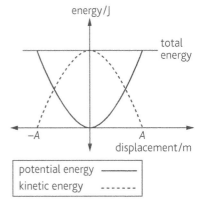

Figure 9.3.4 *Relationship between energy and displacement*

Example

A spring is hung from a fixed point. A mass of 150 g is hung from the free end of a spring. The mass is pulled downwards from its equilibrium position through a small distance y and is released. The mass undergoes simple harmonic motion. Figure 9.3.5 shows the variation with displacement x from the equilibrium position of the kinetic energy of the mass.

Using the figure:

a Determine the distance y through which the mass was initially displaced.

b Determine the angular frequency.

c Determine the frequency of oscillation.

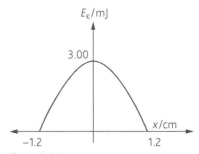

Figure 9.3.5

a $y = 1.2\,\text{cm}$

b Maximum kinetic energy $= 3.00\,\text{mJ}$

Maximum kinetic energy $= \frac{1}{2}mv^2 = \frac{1}{2}m\omega^2(a^2 - x^2)$

$\frac{1}{2} \times 150 \times 10^{-3} \times \omega^2\{(1.2 \times 10^{-2})^2 - 0^2\} = 3.00 \times 10^{-3}$

$\omega = \sqrt{\dfrac{2 \times 3.00 \times 10^{-3}}{150 \times 10^{-3} \times (1.2 \times 10^{-2})^2}} = 16.7\,\text{rad s}^{-1}$

c Frequency of oscillation $f = \dfrac{\omega}{2\pi} = \dfrac{16.7}{2\pi} = 2.66\,\text{Hz}$

Key points

■ To derive an equation for the period of oscillation of a system in simple harmonic motion, the restoring force must first be determined. The acceleration of the mass is then compared with the defining equation for simple harmonic motion.

■ There is a constant interchange between potential and kinetic energy for a system oscillating with simple harmonic motion.

On completion of this section, you should be able to:

- describe practical examples of damped oscillations
- describe practical examples of forced oscillations
- understand the concept of resonance
- identify situations where resonance is useful and when it should be avoided.

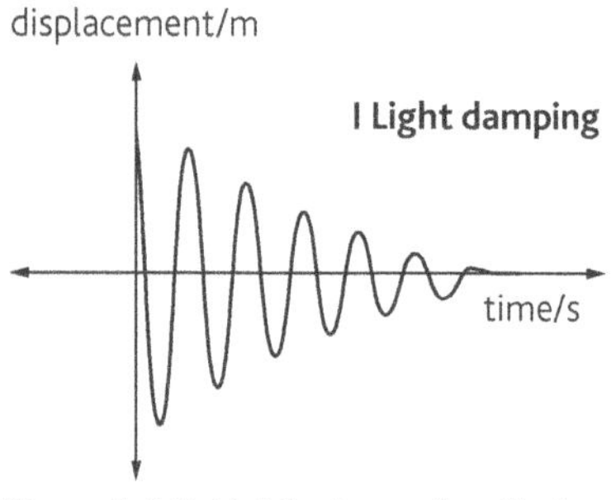

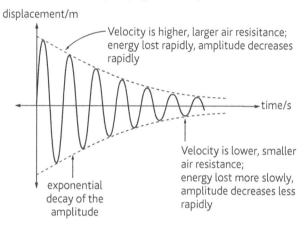

Damped oscillations

When a simple pendulum is displaced slightly and released, it begins to oscillate with simple harmonic motion. The amplitude of the oscillation gradually decreases over time and the pendulum eventually comes to rest at its equilibrium position. Since the pendulum is oscillating in air, air resistance causes energy to be transferred away from the oscillating pendulum. That is to say, work is done against the air resistance. The amplitude decreases in an exponential manner. The oscillation of the pendulum is said to be **damped** (Figure 9.4.1).

displacement/m

Velocity is higher, larger air resistance; energy lost rapidly, amplitude decreases rapidly

time/s

exponential decay of the amplitude

Velocity is lower, smaller air resistance; energy lost more slowly, amplitude decreases less rapidly

Note: the period of oscillation remains constant

Figure 9.4.1 Graph showing damped oscillations

Initially, the pendulum has its maximum energy when it was displaced. This is in the form of potential energy. This energy is converted into kinetic energy as the pendulum bob begins moving back to the equilibrium position. Since the magnitude of the air resistance is dependent on velocity, it will be greatest at the start of the oscillations. The oscillating pendulum will therefore lose energy at a rapid rate. The amplitude of oscillation therefore reduces rapidly. As some time passes, the velocity of the bob is smaller than the initial oscillations. The air resistance is now smaller. Energy will be lost at a slower rate and the amplitude would decrease more slowly. This explains why the amplitude decreases in an exponential manner. It should be noted that even though the motion of the pendulum is damped, the amplitude of the oscillations decreases but the period of oscillation remains constant.

There are different degrees to which a system can be damped. They are:

- **lightly damped** oscillations (e.g. a pendulum oscillating in air)
- **critically damped** oscillations (e.g. a car suspension system)
- **heavily damped** oscillations.

For a system that is lightly damped the amplitude of the oscillation eventually decreases to zero as the system comes to rest (Figure 9.4.2).

For a system that is critically damped, the system comes to rest after one oscillation (Figure 9.4.3).

For a system that is heavily damped, the system fails to oscillate (Figure 9.4.4).

displacement/m

I Light damping

time/s

Figure 9.4.2 Lightly damped oscillations

displacement/m

II Critical damping

time/s

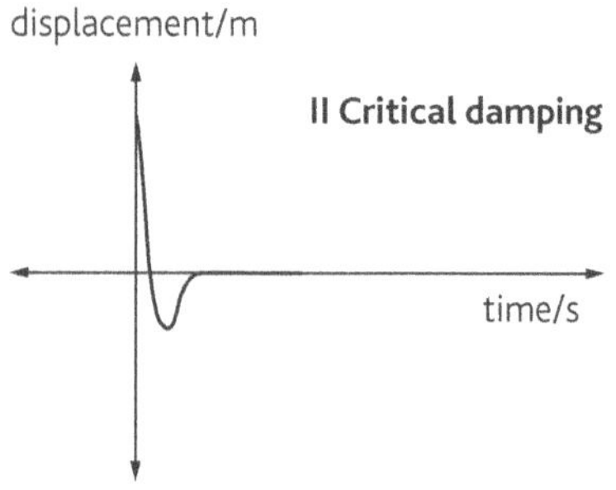

Figure 9.4.3 Critically damped oscillations

displacement/m

III Heavy damping

time/s

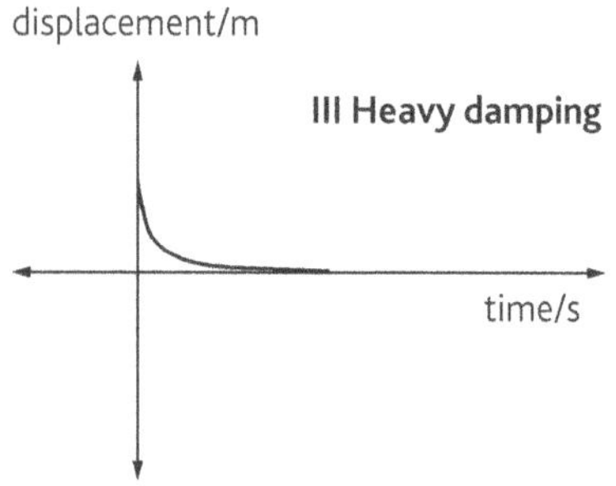

Figure 9.4.4 Heavily damped oscillations

Resonance

Systems that oscillate with simple harmonic motion do so at a particular frequency. This particular frequency is known as the **natural frequency** f_0 of the system. In the case of a simple pendulum of length l, the natural frequency is given by

$$f_0 = \frac{1}{2\pi}\sqrt{\frac{g}{l}}$$

In the case of mass m attached to a spring, having a spring constant k, the natural frequency is given by

$$f_0 = \frac{1}{2\pi}\sqrt{\frac{k}{m}}$$

If a periodic force is applied to a system such that it forces it to oscillate, the amplitude of vibration increases significantly when the frequency of the periodic force is equal to the natural frequency of the system. This phenomenon is known as **resonance**. At resonance, energy is transferred to the system by the periodic force. The periodic force is sometimes referred to as the driver.

Figure 9.4.5 illustrates how the amplitude of an oscillating system varies as the frequency of the periodic force changes. The amplitude of the oscillations is at a maximum when the frequency of the periodic force is equal to the natural frequency of the system.

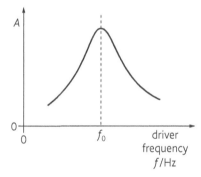

Figure 9.4.5 *Resonance*

A simple laboratory experiment can be set up to demonstrate the phenomenon of resonance. The arrangement is referred to as Barton's pendulums. A string AOB is set up as shown in Figure 9.4.6. The ends are fixed at A and at B. The arrangement consists of several pendulums of varying lengths attached to OB. Each pendulum will have a different period of oscillation. The pendulum OD has a mass attached to it and is called the driver pendulum. The pendulums P, Q, R and S have small inverted paper cones attached to them. The driver pendulum is displaced slightly and begins oscillating. Since all the pendulums are attached to the string OB, P, Q, R and S begin oscillating. The pendulum having a length similar to the driver pendulum, builds up a larger amplitude than the rest of them.

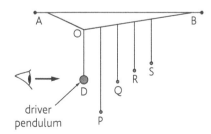

Figure 9.4.6 *Investigating resonance*

Experiment to investigate the effect of damping

A mass is attached between two springs as shown in Figure 9.4.7.

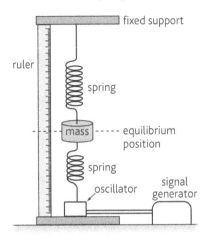

Figure 9.4.7 *Experiment to demonstrate resonance*

A ruler is placed adjacent to the mass so that displacements can be measured. The equilibrium position is first noted on the ruler. An oscillator is connected to a signal generator. The mass is displaced slightly. The amplitude of the oscillation is measured from the metre rule. The period of the oscillation is determined using a stop watch.

The frequency $f_0 = \frac{1}{2\pi}\sqrt{\frac{g}{l}}$ is then determined. The signal generator is then turned on. The signal generator is set to produce a sinusoidal signal. This signal causes the oscillator to move the spring and forces the mass to oscillate. The frequency of the signal generator is varied and the amplitude of oscillation of the mass is recorded. The frequency of the signal generator is measured using a cathode ray oscilloscope.

Figure 9.4.8 shows the results of the experiment.

The same experiment can be used to show the effect of damping on the resonance curve. A small card is attached to the mass as shown in Figure 9.4.9.

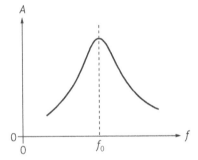

Figure 9.4.8 *Results of experiment*

The card damps the motion of the mass by increasing the drag. The experiment is repeated and the resonance curve is plotted. Figure 9.4.10 shows the effect of damping on the resonant curve. There are two things that are important to note. The peak of the curve is flatter and wider and the resonant frequency is also lower.

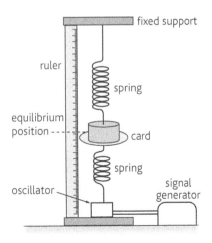

Figure 9.4.9 *Use of a card to damp the oscillations*

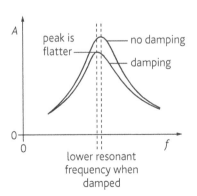

Figure 9.4.10 *Effect of damping on the resonant curve*

Unwanted problems associated with resonance

In cities, tall buildings oscillate naturally. In areas where earthquakes are common, large amounts of destruction can occur. The waves produced by earthquakes forces buildings to oscillate. Large amounts of energy are transferred to buildings. As resonance occurs, buildings are destroyed in seconds. In 2010, an earthquake of magnitude 7.0 caused major destruction in Haiti. In wealthy countries, buildings are designed with dampers to reduce the effects of earthquakes.

Situations where resonance is useful

Microwave cookers

Even though the effects of resonance can be disastrous, there are many examples where resonance is useful. A common household appliance is the microwave. Microwave ovens make use of resonance. Water molecules oscillate at a frequency that lies in the microwave region of the electromagnetic region. Microwave cookers exploit this fact. The microwaves force the water molecules inside the food to oscillate. The temperature of the water increases and thermal energy spreads throughout the food, thereby warming it.

Magnetic resonance imaging (MRI)

Magnetic resonance imaging is a non–invasive medical diagnostic technique used to view internal structures and processes occurring in the human body. In the human body there are huge amounts of hydrogen nuclei present. Hydrogen nuclei are used as the basis for this imaging technique. The patient is placed in a large magnetic field. Radio frequency pulses are transmitted to the person and cause resonance to occur. The hydrogen nuclei emit radio frequency signals which are detected and processed to produce an image.

Electric circuits

Tuning circuits in electrical devices such as radios, make use of resonance. These types of circuits have reactive elements such as capacitors and inductors. An electric current will oscillate between the two components at the circuit's resonant frequency.

Key points

- Air resistance and friction cause oscillations to be damped.
- The amplitude of oscillations decreases when damped.
- Oscillations can be lightly, critically or heavily damped.
- Resonance occurs when the frequency of the driver is equal to the natural frequency of the oscillating system.
- Damping affects a resonant frequency curve.
- There are situations where resonance can be useful and situations where the effects can be catastrophic.

10 Refraction

10.1 Refraction

Figure 10.1.1 *A container of water*

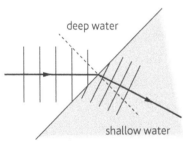

Note:
- speed and wavelength decreases
- frequency is unchanged
- direction of wavefronts change

Figure 10.1.2 *Refraction of water waves*

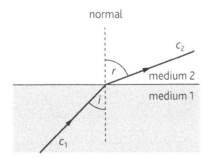

Figure 10.1.3

72

Refraction

Consider a large container of water. A piece of wood is placed on one side of the container to create a deep water section and a shallow water section (Figure 10.1.1). A straight bar is then used to produce straight wavefronts travelling from the deep water to the shallow water (Figure 10.1.2). If the wavefronts strike the deep water–shallow water boundary at some angle, the wavefronts change direction. This change in direction occurs as a result of a change in speed of the wave and is called **refraction**. The wavelength of the wave also changes. The *frequency remains constant* as the wave travels from deep water to shallow water.

Refraction of light

Light behaves as a wave and as such can be refracted. Refraction occurs when a ray of light travels between two media of different optical densities. Consider a ray of light travelling from medium 1 to medium 2 as shown in Figure 10.1.3. The **normal** is a line drawn at right angles to the surface where the ray strikes the boundary between the two media. The **angle of incidence** i is the angle between the normal and the incident ray. The **angle of refraction** r is the angle between the normal and the refracted ray.

The ratio $\dfrac{\sin i}{\sin r}$ is a constant, and is called the **refractive index**.

The refractive index of medium 2 with respect to medium 1 is $_1n_2$.

Therefore, $_1n_2 = \dfrac{\sin i}{\sin r}$. This is known as **Snell's law**.

It can also be shown that $_1n_2 = \dfrac{c_1}{c_2}$, where c_1 is the speed of light in medium 1 and c_2 is the speed of light in medium 2.

If the ray of light was travelling from medium 2 to medium 1, the refractive index of medium 1 with respect to medium 2 is $_2n_1$.

Therefore, $_2n_1 = \dfrac{1}{_1n_2}$

The **absolute refractive index,** n, of a material is defined as $n = \dfrac{c_v}{c_m}$, where c_v is the speed of light in a vacuum and c_m is the speed of light in the material.

Suppose the absolute refractive index of medium 1 and medium 2 were n_1 and n_2 respectively.

For medium 1,	$n_1 = \dfrac{c_v}{c_1}$	Equation (1)
For medium 2,	$n_2 = \dfrac{c_v}{c_2}$	Equation (2)

Equation (2) ÷ Equation (1) $\quad \dfrac{n_1}{n_2} = \dfrac{c_v}{c_2} \div \dfrac{c_v}{c_1} = \dfrac{c_v}{c_2} \times \dfrac{c_1}{c_v} = \dfrac{c_1}{c_2}$

Therefore, Snell's law can be rewritten as $\dfrac{\sin i}{\sin r} = \dfrac{n_1}{n_2}$ or $n_1 \sin i = n_2 \sin r$

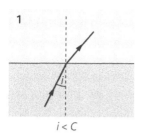

$i < C$

Laws of refraction

1 The incident ray, the refracted ray and the normal at the point of incidence lie in the same plane.

2 The ratio $\dfrac{\sin i}{\sin r}$ is constant, where i is the angle of incidence and r is the angle of refraction.

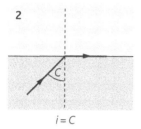

$i = C$

When a ray of light travels from a medium of lower optical density, to a medium of higher optical density, the refracted ray bends towards the normal (e.g. air to glass).

When a ray of light travels from a medium of higher optical density, to a medium of lower optical density, the refracted ray bends away from the normal (e.g. glass to air).

Critical angle and total internal reflection

Consider a ray of light travelling from glass to air. There is an angle of incidence called the **critical angle, *C*,** for which the angle of refraction is 90°. There are three scenarios to consider (Figure 10.1.4).

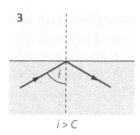

$i > C$

1 The angle of incidence $< C$. The ray is refracted out of the glass.

2 The angle of incidence $= C$. The angle of refraction is 90° and the ray is refracted along the surface of the glass.

3 The angle of incidence $> C$. The ray is reflected back inside the glass.

$n_1 \sin 90° = n_2 \sin C$, but $\sin 90° = 1$

Figure 10.1.4

Therefore, $\sin C = \dfrac{n_g}{n_a} = \dfrac{1}{\,_g n_a}$

In scenario 3, the ray of light is said to be totally internally reflected. The angle of incidence and the **angle of reflection** are equal.

Conditions for **total internal reflection**

1 Light must be travelling from an optically dense to an optically less dense medium (e.g. glass to air).

2 The angle on incidence must be greater than the critical angle.

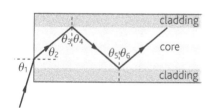

Figure 10.1.5 *A fibre optic cable*

Applications of total internal reflection

Total internal reflection is used in fibre optic cables, very thin glass fibres surrounded by a material called cladding. The cladding has a lower optical density than glass. A ray of light, typically a laser, is projected into the glass fibre at one end. The angle of incidence inside the fibre is greater than the critical angle for the glass–cladding boundary and total internal reflection occurs. The light is guided along the fibre until it reaches the other end. Fibre optic cables are used extensively in the field of communications to transmit data in the form of light pulses (Figure 10.1.5).

Expensive engagement rings contain diamond crystals which 'sparkle'. Cheaper rings made from glass sparkle much less, because glass has a refractive index of 1.50, while diamond has a larger refractive index of 2.42. The critical angles for glass and diamond are 41.8° and 24° respectively. Each ray of light that enters diamond reflects many times from the flat surfaces inside the diamond before finally emerging. The many reflections are perceived by our eyes as 'sparkle'. The smaller critical angle of diamond means that there is a greater chance for many total internal reflections, hence more 'sparkle' than in glass.

Key points

- Refraction is the change in direction of a wave that occurs as a result of a change in speed of the wave.

- The ratio $\dfrac{\sin i}{\sin r}$ is a constant, and is called the refractive index.

- The critical angle, C, is the angle of incidence for which the angle of refraction is 90°.

- Total internal reflection occurs when the angle of incidence is greater than the critical angle.

Revision questions 4

Answers to questions that require calculation can be found on the accompanying CD.

1 Explain what is meant by:
   a oscillations [1]
   b free oscillations [1]
   c simple harmonic motion. [2]

2 Describe an example of a free oscillation. [2]

3 The centre of a cone of a loudspeaker is oscillating with simple harmonic motion of frequency 1200 Hz and amplitude 0.07 mm. Calculate
   a the angular frequency of the oscillations [2]
   b the maximum acceleration of the centre of the cone. [2]

   Sketch a graph to show the variation with displacement x of the acceleration of the centre of the cone. [3]

4 A pendulum bob oscillates with simple harmonic motion. Its displacement varies with time as shown below.

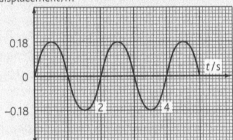

Determine:
   a the amplitude of the oscillation [1]
   b the period of the oscillation [1]
   c the frequency [1]
   d the angular frequency [1]
   e the acceleration [1]
      i   when the displacement is zero [1]
      ii when the displacement is at a maximum [2]
   f the maximum velocity of the pendulum bob. [2]

5 Derive an expression for the period of oscillation of a simple pendulum. [6]

6 Derive an expression for the period of oscillation of two identical springs in parallel. [6]

7 a Calculate the gain in potential energy when a mass of 120 g is raised through 1.1 mm. [2]
   b A simple pendulum consists of a light inextensible string and a bob of mass of 120 g. The variation of the potential energy with x, the horizontal displacement of the bob is shown below.

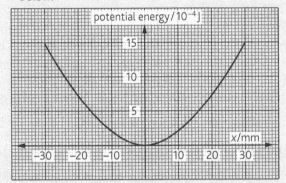

   The bob of the pendulum is displaced sideways until its centre of mass is raised through a vertical distance of 1.1 mm and then released.
   c Copy the figure and sketch graphs to show the variation, as the pendulum oscillates, of x with:
      i   the total energy [2]
      ii the kinetic energy. [2]
   d Determine the amplitude of oscillation of the pendulum. [2]

8 a Explain what is meant by the term 'resonance'. [2]
   b A piece of Plasticine® is attached to one end of a spring. The spring is attached to a support that is able to vibrate in a vertical plane. The support begins to vibrate and the mass-spring system is forced to oscillate.
      i   Explain what is meant by forced oscillations. [2]
      ii Sketch a graph to show the variation of amplitude of the mass with frequency of vibration of the support. [3]
      iii The Plasticine® is now flattened so that it causes the oscillations to be damped. On the same axes as in b (ii) sketch another graph to show the effect of damping. [3]
   c State one situation in which resonance is useful. [1]
   d State one situation in which resonance can be a hazard. [1]

9 a State the laws of refraction. [2]

b The speed of light in air is 3.00×10^8 m s^{-1}. The speed of light in glass is 1.99×10^8 m s^{-1}. Consider a ray incident on a face of a prism as shown below.

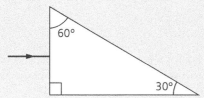

i Calculate the refractive index of the glass used to make the prism. [2]

ii Calculate the critical angle for a glass/air interface. [2]

iii Sketch a diagram to show what happens to the ray inside the prism and as it leaves the prism. [3]

iv Calculate the angle of refraction when the ray emerges from the prism. [4]

10 Discuss one application of total internal reflection. [2]

11 The refractive index of glass for red light is 1.510. Red light travelling at 3.00×10^8 m s^{-1} is incident at an angle of 32° on an air–glass boundary. Calculate:

a the angle of refraction for red light

b the speed of red light in glass

c the critical angle for the air–glass interface.

11.1 Waves

Learning outcomes

On completion of this section, you should be able to:

- understand that a wave transmits energy

- define the terms *displacement*, *amplitude*, *period*, *frequency* and *speed* when applied to a wave.

- understand the terms *phase* and *phase difference*.

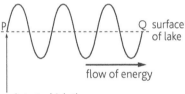

Figure 11.1.1 *Profile of a wave*

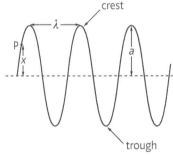

Figure 11.1.2 *Snapshot of a wave*

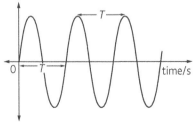

Figure 11.1.3 *Variation of displacement of P with time*

Describing a wave

When a stone is dropped into a lake at a point P, circular ripples are seen moving away from P. The ripples or **waves** transmit energy away from the point of impact on the surface of the lake. The point Q is a short distance away from P. A side profile from the point P to Q is shown in Figure 11.1.1.

It is important to understand that the water wave transfers energy from the point P to Q, without any water particles actually travelling from P to Q. The water particles along the line PQ oscillate about a fixed point. It is the energy that is transmitted by the wave. There are various ways of classifying waves. A **mechanical wave** is one that requires a substance through which to propagate. Mechanical waves can be classified as either **transverse** or **longitudinal waves**. In order to describe a wave, a transverse wave will be considered. The wave produced in the lake when the stone was dropped into it is an example of a transverse wave. The particles that make up this type of wave oscillate at right angles to the direction of propagation of the wave.

Another way to produce a transverse wave is to fix one end of a rope and move the free end up and down with your hand until waves are produced. Viewing the wave from the side will produce a picture similar to the wave in the lake.

Figure 11.1.2 illustrates the main characteristics of the wave.

The dashed line represents the rope in its undisturbed or rest position. The up and down motion of the hand generates a transverse **progressive wave**. A progressive wave is one that moves in a particular direction, carrying energy along with it. The wave in Figure 11.1.2 is made up of a series of crests and troughs. The distance x, represents the **displacement** of the point P on the wave from the rest position. The SI unit is the metre.

The **wavelength** λ of a wave is the distance between two successive crests or two successive troughs. The SI unit is the metre.

The **amplitude** a of a progressive wave is the maximum displacement from the rest position. The SI unit is the metre.

Suppose that the movement of a single particle (point P) is observed over a period of time. The point P oscillates about the rest or equilibrium position of the rope. Figure 11.1.3 shows how the displacement of the particle varies with time. The graph shows that the displacement of the particle varies sinusoidally with time.

One complete cycle or oscillation is made up of one crest and one trough. The time taken for one cycle or oscillation is the called the **period** of oscillation of the wave and is shown by T in Figure 11.1.3. The SI unit is the second.

The number of oscillations per unit time of a point on a wave is known as the **frequency** of the wave and is measured in hertz (Hz).

The frequency and period of a wave are related by the following equation.

$$T = \frac{1}{f}$$

T – period/s;   f – frequency/Hz

The **speed** of a wave is the distance travelled per unit time. Speed in this case refers to the rate at which energy is being transferred. The frequency and the wavelength can be used to determine the velocity of a wave. The derivation is shown below.

Time taken for one oscillation $= T$

During this time the wave would travel a distance $= \lambda$

$$\text{speed} = \frac{\text{distance}}{\text{time}}$$

$$v = \lambda \times \frac{1}{T}$$

$$= \frac{\lambda}{T}$$

But $\qquad T = \frac{1}{f}$

$\therefore \qquad v = f\lambda$

v – speed/m s^{-1}; $\qquad f$ – frequency/Hz; $\qquad \lambda$ – wavelength/m

Phase and phase difference

All the particles in a mechanical wave vibrate about their mean positions. Not all the particles that make up the wave move together. At any particular instant in time, some particles may be moving upwards, while some particles may be moving downwards.

For example, the particles P and Q on the transverse wave in Figure 11.1.4 are both moving upwards at the same time. P and Q are said to be in phase with each other. The particle at the point R is moving downward and is therefore out of phase with particle Q. The **phase** of a particular point on a wave is a measure of the fraction of the oscillation that has been completed. Phase relationships are sometimes measured in degrees or radians. Particles that are in phase with each other have zero phase difference. Particles that are completely out of phase have a **phase difference** of π radians or 180°.

Consider two waves P and Q shown in Figure 11.1.5

The two waves are not in phase with each other. A phase difference exists between the two waves. The phase difference ϕ between them is determined by the following:

Phase difference $\quad \phi = \frac{y}{\lambda} \times 2\pi$

The graph in Figure 11.1.6 shows two waves P and Q that completely out of phase. The phase difference between them is π radians or 180°.

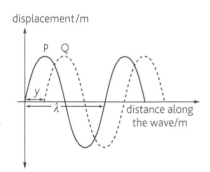

Figure 11.1.5 Phase difference between two waves

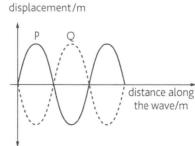

Figure 11.1.6 Waves that are out of phase

☑ *Exam tip*

When reading the wavelength or period from a graph check the label on the *x*-axis.

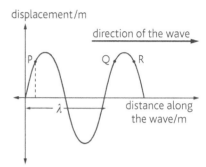

Figure 11.1.4 *The phase of a point on a wave*

Key points

- Waves transmit energy.

- Mechanical waves require a material medium through which to travel.

- Mechanical waves can either be transverse or longitudinal.

- Displacement of a point on a wave is the distance from the equilibrium position.

- The wavelength of a wave is the distance between two successive points in phase.

- The amplitude of a wave is the maximum displacement from the equilibrium position.

- The period of a wave is the time taken for one oscillation.

- The frequency of a wave is the number of oscillations per second.

- The speed of a wave is the rate at which energy is being transferred.

- The phase of a particular point on a wave is a measure of the fraction of the oscillation that has been completed.

11.2 Transverse and longitudinal waves

Learning outcomes

On completion of this section, you should be able to:

- understand the difference between a transverse and a longitudinal wave

- appreciate that a progressive wave transmits energy from one point to another

- understand that the intensity of a progressive wave is proportional to the amplitude squared

- explain the term *polarisation*.

- understand that transverse waves can be polarised and longitudinal waves cannot.

Transverse and longitudinal waves

Mechanical waves can be classified based on the movement of the particles that make up the wave. Two types of mechanical waves are transverse waves and longitudinal waves.

In order to differentiate between the two types of waves, a slinky spring can be used. In order to produce a **transverse wave**, one end of the slinky is fixed and the other end is moved up and down repeatedly. In this type of wave, the particles oscillate at right angles to the direction of travel of the wave. Light is an example of a transverse wave.

In order to produce a **longitudinal wave**, one end of the slinky is fixed and the other end is moved back and forth as shown in Figure 11.2.1. In this type of wave, the particles oscillate in the same direction of travel of the wave. Sound is an example of a longitudinal wave.

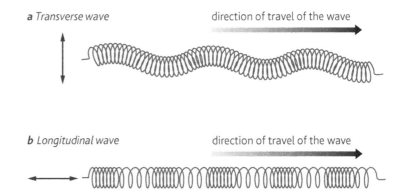

a Transverse wave   direction of travel of the wave

b Longitudinal wave   direction of travel of the wave

Figure 11.2.1 *How a slinky spring can be used to illustrate transverse and longitudinal waves*

A transverse wave is easily illustrated using a sinusoidal curve. In the case of a longitudinal wave, visualising it can be difficult.

When a longitudinal wave is travelling through a medium, it consists of a series of **compressions** and **rarefactions**. A compression is a region in which the particles are moving towards each other. A rarefaction is a region where the particles are moving away from each other.

A longitudinal wave is sometimes illustrated as a transverse wave. However, the compressions and rarefactions correspond to peaks and troughs on a sinusoidal waveform. Figure 11.2.2 illustrates a longitudinal wave produced by a loudspeaker. The sinusoidal wave illustrates the variation of pressure of the air molecules in front of the loudspeaker. At a compression, the air molecules are moving towards each other and this corresponds to a high pressure. At a rarefaction, the air molecules are moving away from each other and this corresponds to a low pressure.

Definitions

In a transverse wave, the particles in the medium vibrate at right angles to the direction of energy transfer.

In a longitudinal wave, the particles in the medium vibrate in the same direction of energy transfer.

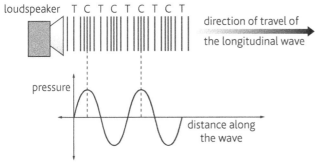

Figure 11.2.2 *Visualising a longitudinal wave*

Progressive waves

A **progressive wave** is a wave that is produced as a result of vibrations and transfers energy from one location to another.

Both transverse and longitudinal waves can be progressive waves. Light is an example of a traverse progressive wave. Sound is an example of a longitudinal progressive wave. As the wave moves in a particular direction, energy is transmitted in that same direction. The intensity of a wave is related to its amplitude.

$$I \propto A^2$$

The intensity of a wave is proportional to the square of the amplitude of the wave. This means that a wave with a large amplitude transmits more energy than a wave with a smaller amplitude. Take for example, the sound waves being produced by a loudspeaker. When the volume is low, the speaker produces longitudinal progressive waves with small amplitudes. This is perceived by the human ear as a low decibel level. These waves transmit a small amount of energy. When the volume is increased, the speaker produces longitudinal sound waves with larger amplitudes. This is perceived by the human ear as a high decibel level. In some instances, you will be able to feel the vibrations from the speaker.

Example

A sound wave of amplitude 0.15 mm has an intensity of 3.2 W m^{-2}. Calculate the intensity of a sound wave of the same frequency, which has an amplitude of 0.45 mm.

Recall that $\qquad I \propto A^2$

$\therefore \qquad\qquad I = kA^2$

Initially, $A = 0.15$ mm and $I = 3.2$ W m^{-2}.

$$3.2 = k \times (0.15 \times 10^{-3})^2$$

$$k = \frac{3.2}{(0.15 \times 10^{-3})^2} = 1.422 \times 10^8$$

Now the frequency of the sound wave is the same, but the amplitude has changed to 0.45 mm.

$$I = kA^2$$

$$= (1.422 \times 10^8) \times (0.45 \times 10^{-3})^2$$

$$= 28.8 \text{ W m}^{-2}$$

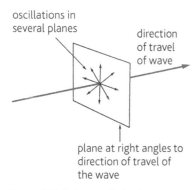

oscillations in several planes

direction of travel of wave

plane at right angles to direction of travel of the wave

Figure 11.2.3 *An unpolarised wave*

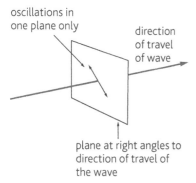

oscillations in one plane only

direction of travel of wave

plane at right angles to direction of travel of the wave

Figure 11.2.4 *A polarised wave*

Polarisation

In transverse waves, the oscillations are perpendicular to the direction of propagation of the wave. In an unpolarised transverse wave, the plane of oscillation can be in any one of an infinite number of planes. This is illustrated in Figure 11.2.3.

In a polarised wave, the oscillations are restricted to one plane as shown in Figure 11.2.4.

An easy way to understand the concept of polarisation is to fix one end of a rope and move the free end up and down with your hand until waves are produced. The waves produced are in a vertical plane. These waves are said to be vertically polarised. If you move your hand horizontally from side to side, the waves produced lie in a horizontal plane. These waves are said to be horizontally polarised. It is possible to produce waves in different planes of polarisation by simply adjusting the angle at which your hand moves.

Consider a situation where the light from a filament lamp is viewed with the naked eye. An intensity, I, will be observed. When a sheet of Polaroid® is placed in front of the lamp and then viewed, the intensity of the light appears to have reduced. The reason for this is that the sheet restricts the oscillations of the light waves to one plane. It therefore prevents most of the light waves from reaching the eye. If the sheet of Polaroid® is now rotated in a plane perpendicular to the direction of travel of light, the intensity is unchanged (Figure 11.2.5).

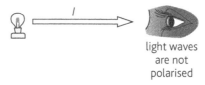

light waves are not polarised

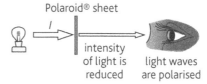

Polaroid® sheet

intensity of light is reduced

light waves are polarised

Figure 11.2.5 *The effect of a sheet of Polaroid®*

Suppose a second sheet of Polaroid® now placed in front of the first, but rotated through 90° to the first sheet. The intensity of the light is now reduced to zero. When the light waves pass through the first sheet, only one plane of light is allowed to pass through. Since the second sheet is rotated through 90°, it will not allow these light waves to pass through and hence the intensity is reduced to zero (Figure 11.2.6).

Polarisation is a property that is only exhibited by transverse waves. Transverse waves have oscillations which are perpendicular to the direction of propagation. In a transverse wave, the oscillations can be in different planes. Polarisation restricts the oscillation to one plane. Longitudinal waves have oscillations parallel to the direction of propagation. Longitudinal waves cannot be polarised. Therefore, **polarisation is a test used to distinguish between transverse and longitudinal waves.**

There are many examples of the polarising effect of light. Polaroid®
sunglasses reduce glare by limiting the amount of light entering the eye.
Liquid crystal displays found in digital watches are polarised. Structural
engineers often use the effect of polarisation to perform experiments
to measure the amount of stress distributed throughout a component
under test.

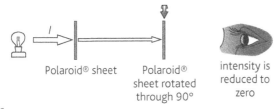

Figure 11.2.6

Table 11.2.1 summarises the similarities and differences between a
transverse wave and a longitudinal wave.

Table 11.2.1 Comparing transverse and longitudinal waves

	Transverse wave	**Longitudinal wave**
Similarities	Transfers energy in the direction of propagation of the wave	Transfers energy in the direction of propagation of the wave
	Shows reflection, refraction, diffraction and interference	Shows reflection, refraction, diffraction and interference
Differences	Particles that make up the wave oscillate at right angles to direction of energy transfer	Particles that make up the wave oscillate in the same direction of energy transfer
	Can be polarised	Cannot be polarised
Examples	Light, microwaves	Sound

Key points

- In a transverse wave, the particles in the medium vibrate at right angles to the direction of energy transfer.
- In a progressive wave, the particles in the medium vibrate in the same direction as energy transfer.
- A progressive wave transfers energy from one point to another.
- The intensity of a progressive wave is proportional to the amplitude squared.
- In an unpolarised wave, the oscillations are in various planes.
- In a plane polarised wave, the oscillations are restricted to one plane.
- Transverse waves can be polarised.
- Longitudinal waves cannot be polarised.

Learning outcomes

On completion of this section, you should be able to:

- state the principle of superposition
- understand the term *diffraction*
- draw diagrams to show narrow and wide gap diffraction
- recall the formula to determine wavelength using a diffraction grating.

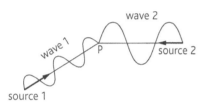

Figure 11.3.1

Definition

Principle of superposition – when two or more waves arrive at a point, the resultant displacement at that point is the algebraic sum of the individual displacements of each wave.

The principle of superposition

When two particles are projected towards each other, they collide and move off in different directions. It is easy to visualise this if you project two steel marbles towards a point P on a table.

The situation is very different with two waves. When two waves are directed towards a point P in space, they pass through each other. However, something interesting happens at the point P. Suppose the amplitude of one of the waves is A and the amplitude of the second wave is $4A$. Assuming that the two waves arrive at the point P in phase with each other, the amplitude of the resultant wave at the point P is $5A$. If the waves arrive at the point P completely out of phase with each other, the amplitude of the resultant wave at P is $3A$. This phenomenon can be explained using the principle of **superposition** (Figure 11.3.1).

The principle of superposition states that when two or move waves arrive at a point, they superimpose on each other and the resultant displacement is the algebraic sum of the individual displacements of each wave.

Consider two waves, wave 1 and wave 2 as shown in Figure 11.3.2. Both waves are in phase with each other. (Phase difference is zero.) When these two waves combine, by applying the principle of superposition, wave 3 is produced. Notice that the amplitude of the resultant wave is twice as large as wave 1 and wave 2. Figure 11.3.3 shows what happens when wave 1 and wave 2 arrive out of phase.

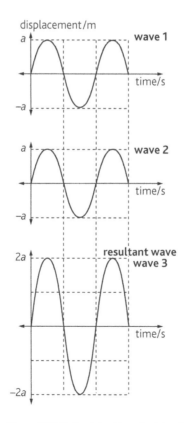

Figure 11.3.2 *Applying the principle of superposition (waves in phase)*

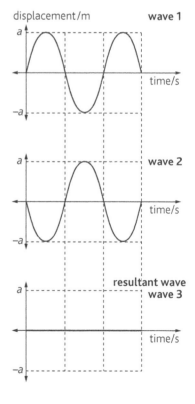

Figure 11.3.3 *Applying the principle of superposition (waves out of phase)*

Diffraction

Diffraction is the spreading out of wavefronts, when a wave passes the edge of an object or through a gap. In order for significant diffraction to be observed through a gap, the wavelength of the wave must be comparable to the width of the gap. It is important to note that there is no change in the wavelength of the wave when diffraction occurs. This information should be illustrated when sketching diffraction diagrams. Light has a very small wavelength. This means that in order for diffraction of light to occur, the width of the gap should be comparable to the wavelength. For this reason the diffraction of light is not normally observed in our everyday experiences. Sound waves on the other hand, have much larger wavelengths. They can be easily diffracted through a doorway. This explains why a CD player can be heard in another room but not necessarily be seen. The light doesn't bend around corners to allow the CD player to be seen.

Diffraction can be easily demonstrated in a ripple tank. Plane waves are generated using a straight bar. As the bar vibrates on the surface of the tank, plane waves are produced and move towards a gap. Figure 11.3.4 (a), (b) and (c) shows what is seen from above the ripple tank when the experiment is performed using gaps and the edge of an object.

Definition

Diffraction is the spreading out of wavefronts of a wave when they pass through a gap or pass the edge of an obstacle.

Figure 11.3.4 (a) Narrow gap diffraction

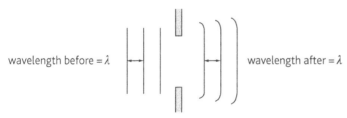

Figure 11.3.4 (b) Wide gap diffraction

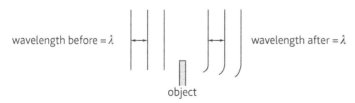

Figure 11.3.4 (c) Diffraction at the edge of an object

If the gap is made much smaller than the wavelength of the wave, no diffraction will occur. The waves will not pass through the gap. It is this same principle that applies to a microwave cooker. The metal grid on the door allows light to pass through so that you can view the food, but doesn't allow the microwaves to escape from inside the microwave.

Single slit diffraction

When a single slit is placed in front of a parallel beam of light, diffraction occurs. The intensity distribution is shown in Figure 11.3.5.

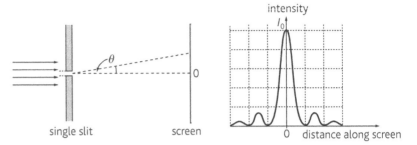

Figure 11.3.5 (a) Single slit diffraction

Figure 11.3.5 (b) Variation of intensity with distance along screen

Diffraction gratings

A **diffraction grating** consists of a large number of equally spaced lines ruled on a piece of glass or plastic. When monochromatic light (a single wavelength e.g. red light) is incident at right angles to the grating, an interference pattern of bright and dark fringes are observed on a screen.

The bright fringes are called maxima. The fringe at the centre is called the zeroth order maximum ($n = 0$). The next fringe is called the first order maximum ($n = 1$). The fringe pattern is symmetrical about the zeroth order maximum. Figure 11.3.6 illustrates how a diffraction grating produces a fringe pattern.

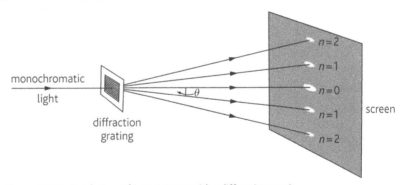

Figure 11.3.6 Producing a fringe pattern with a diffraction grating

Figure 11.3.7 shows the variation of intensity along the fringe pattern.

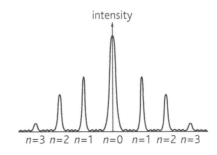

Figure 11.3.7 Variation of intensity along the fringe pattern

The light incident on the grating diffracts at every slit, producing coherent sources of waves. At regions where bright fringes are observed, the path difference from the slits is a whole number of half wavelengths. The light waves are in phase and constructive interference occurs. At regions where dark fringes are observed, the path difference from the slits is an odd number of half wavelengths. The light waves are completely out of phase and destructive interference occurs. The wavelength of light incident on the diffraction grating is related to the angular deviation as follows:

Equation

$d \sin\theta = n\lambda$

d – separation of slits in grating/m
θ – angle between the zeroth order fringe and the nth order fringe
λ – wavelength of light/m
n – nth order fringe

Example

When monochromatic light of wavelength λ is incident normally on a plane diffraction grating, the second order diffraction lines are formed at angles of 30° to the normal to the grating. The diffraction grating has 500 lines per millimetre. Calculate the value of λ.

$$\text{Separation of slits} \quad d = \frac{1 \times 10^{-3}}{500} = 2 \times 10^{-6}\,\text{m}$$

$$d \sin\theta = n\lambda$$

$$(2 \times 10^{-6})(\sin 30°) = 2 \times \lambda$$

$$\lambda = \frac{(2 \times 10^{-6})(\sin 30°)}{2}$$

$$\lambda = 5 \times 10^{-7}\,\text{m}$$

Example

Determine the highest order of diffracted beam that can be produced when a grating with a spacing of 2×10^{-6} m is illuminated normally with light of wavelength 640 nm.

$$d \sin\theta = n\lambda$$

Recall, $\sin\theta \leq 1$ (i.e. the sine of an angle cannot exceed 1)

$$\therefore \quad \frac{n\lambda}{d} \leq 1$$

$$\therefore \quad \frac{n \times (640 \times 10^{-9})}{2.5 \times 10^{-6}} \leq 1$$

$$n \leq \frac{2.5 \times 10^{-6}}{640 \times 10^{-9}}$$

$$n \leq 3.91$$

Since n has to be an integer, the highest order possible is $n = 3$.

This corresponds to 7 principal maxima (the zero order maxima, plus the three on either side).

Exam tip

When white light passes through a diffraction grating, a central bright fringe is observed. On either side of the central bright fringe, a series of spectra (red → violet) are seen.

Key points

- Principle of superposition – when two or more waves arrive at a point, the resultant displacement at that point is the algebraic sum of the individual displacements of each wave.

- Diffraction is the spreading out of wavefronts of a wave when they pass through a gap or pass the edge of an obstacle.

Learning outcomes

On completion of this section, you should be able to:

- explain what is meant by interference

- understand the terms *coherence* and *path difference*

- explain Young's double slit experiment using the principle of superposition

- state the conditions necessary to produce an observable interference pattern.

Young's double slit experiment

Thomas Young performed an experiment that showed that light was a wave and produces an interference pattern. Monochromatic light is made to shine on a card with two slits in it. The slits are a fraction of a millimetre in width and are about one millimetre apart. A screen is set up about one metre or more away from the slits (Figure 11.4.1). A series of equally spaced bright and dark fringes is observed on the screen (Figure 11.4.2). Figure 11.4.3 shows the variation of the intensity of light along the fringe pattern.

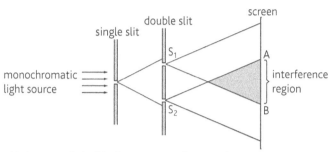

Figure 11.4.1 Young's double slit experiment (Top view)

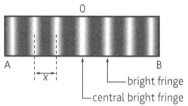

Figure 11.4.2 Fringe pattern produced on the screen

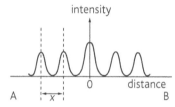

Figure 11.4.3 Variation of intensity of light along the fringe pattern

Coherent waves

Coherent waves have the same frequency and hence a constant phase difference between them. Consider two waves P and Q. If P and Q are in phase with each other, the phase difference between them is zero. P and Q are said to be coherent waves. If P and Q are completely out of phase with each other, the phase difference between them is 180°. P and Q are still coherent waves.

To produce coherent light sources, a single slit is used. As the light passes through the double slit, diffraction takes place and the light waves overlap.

Explaining the experiment (light as a wave)

At points where the crests of waves coincide, **constructive interference** occurs (Figure 11.4.4). These points correspond to bright fringes on the screen. At points where crests and troughs coincide, **destructive interference** occurs (Figure 11.4.5). These points correspond to dark fringes on the screen. In both cases, the waves are superposed.

Path difference

Figure 11.4.6 illustrates two slits, S_1 and S_2. The distance between them is a. The central bright fringe is at O and the first bright fringe occurs at the point P.

Definition

Two waves are said to be coherent if there is a constant phase difference between them.

Definitions

Constructive interference occurs when the waves arrive in phase (crests coincide). The resultant displacement is greater than either of the two waves.

Destructive interference occurs when the waves arrive out of phase (crest and trough coincide). The resultant displacement is zero.

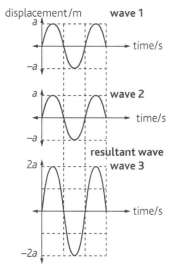

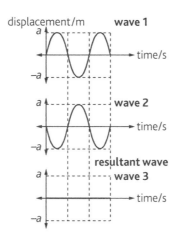

Figure 11.4.4 *Constructive interference*

Figure 11.4.5 *Destructive interference*

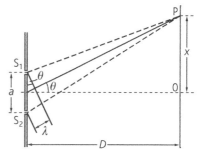

Figure 11.4.6 *Understanding path difference*

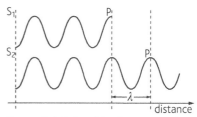

Figure 11.4.7 *Condition for constructive interference*

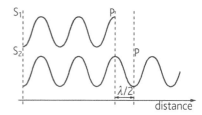

Figure 11.4.8 *Condition for destructive interference*

Suppose S_1 and S_2 emit wave crests at the same time. The point P is a bright fringe and therefore constructive interference must occur at this point. The distance S_2P is greater than S_1P. Therefore, the wave crest from S_2 has to travel a longer distance to arrive at P when compared to the wave crest from S_1. In order for constructive interference to occur at P, the two crests must coincide. This can only happen if the extra distance travelled by the wave from S_2 is equal to exactly one wavelength λ. This extra distance is called the **path difference** (Figure 11.4.7).

For constructive interference to occur at P, the path difference must be zero or a whole number of wavelengths.

$$S_2P - S_1P = n\lambda \qquad n = 0, 1, 2, \ldots$$

So the path difference must be 0, λ, 2λ, …

Where P is the first dark fringe, destructive interference must occur at P. For destructive interference to occur, a crest and a trough must coincide. This can only happen if the extra distance travelled by the wave from S_2 is equal to exactly $\frac{1}{2}\lambda$ (Figure 11.4.8).

For destructive interference to occur at P, the path difference must be an odd number of half-wavelengths.

$$S_2P - S_1P = \left(n + \frac{1}{2}\right)\lambda \qquad n = 0, 1, 2, \ldots$$

So the path difference must be $\frac{1}{2}\lambda$, $\frac{3}{2}\lambda$, $\frac{5}{2}\lambda$, …

Conditions for interference to take place

Under normal conditions, the interference of light is not easily observed. There are conditions that must be satisfied to produce observable interference patterns:

- The waves must be coherent.
- The waves must meet at a point.
- If the waves are polarised, they must be in the same plane of polarisation.
- The waves must be of the same type.
- The amplitudes of the waves must be similar.
- The waves must have the same frequency.

Key points

- Young's double slit experiment shows that light behaves as a wave.

- The principle of superposition can be used to explain the experiment.

- Constructive interference occurs when waves arrive in phase with each other.

- Destructive interference occurs when waves arrive out of phase with each other.

- Coherent waves are waves that have a constant phase difference between them.

- Observable interference patterns can only be observed only when certain conditions are met.

11.5 Interference experiments

Learning outcomes

On completion of this section, you should be able to:

- describe experiments to demonstrate interference of water waves, sound waves and microwaves.

Demonstrating interference in a ripple tank

A straight piece of wood can be used to produce plane wave fronts that travel towards two gaps as shown in Figure 11.5.1. Diffraction occurs at each of the gaps. This causes the waves to overlap in the region beyond the openings. As a result, an interference pattern is produced. There are points on the surface of the water where it is stationary, and points where it is disturbed. If we assume that the wavefronts represent crests, then points of intersection would be points at which constructive interference occurs. This corresponds to points of intersection of the line AB with the curved wavefronts. At points where crests and troughs meet, destructive interference occurs. This corresponds to the points of intersection of the line CD with the curved wavefronts.

The experiment can also be performed using two dippers attached to the same vibrating strip (Figure 11.5.2). This would ensure that the sources of waves are coherent. The interference pattern would be similar to that produced in Figure 11.5.1.

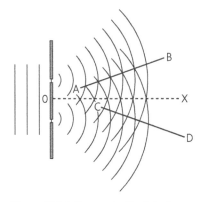

Figure 11.5.1 *Demonstrating interference in a ripple tank*

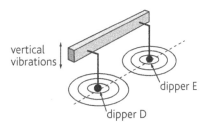

Figure 11.5.2 *Using dippers to produce circular wavefronts*

Demonstrating the interference of sound waves

In order to demonstrate the interference of sound waves the following are required:

- two identical loudspeakers
- a signal generator.

The two loudspeakers are connected to a signal generator. The loudspeakers are adjusted so that they are about 0.5 to 1.0 m apart and facing the same direction (Figure 11.5.3). The signal generator is adjusted so that it produces a signal of frequency in the range 500 Hz to 2 kHz. A microphone is connected to an oscilloscope. An observer walks along the line XY in front of the loudspeakers. The observer will notice a variation in loudness along the line XY. At the point O, sound is heard (large amplitude on the oscilloscope). The path difference in this case is zero and constructive interference occurs at O. When the microphone is moved from O towards Y, the loudness will reach a minimum value (small amplitude on the oscilloscope). At a point of minimum loudness, the sound waves are out of phase and destructive interference occurs.

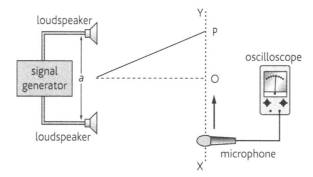

Figure 11.5.3 *Demonstrating interference of sound waves*

Demonstrating the interference of microwaves

In order to demonstrate the interference of microwaves the following are required:

- a microwave transmitter
- a metal plate with two slits
- a microwave detector.

Microwaves from the transmitter T are incident on the two slits S_1 and S_2 which are equidistant from T. S_1 and S_2 act as coherent sources of microwaves. The waves are diffracted by the slits and superpose in the region beyond it. A microwave detector is moved along the line AB, which is parallel to the plane of the slits (Figure 11.5.4). The intensity measured at O is a maximum. The path difference in this case is zero and constructive interference occurs at O. When the microwave detector is moved from O towards A, the intensity decreases to a minimum before increasing to another maximum at the point X. At a point of minimum intensity, the microwaves are out of phase and destructive interference occurs. In this demonstration, the distance between the slits is approximately 1–8 cm.

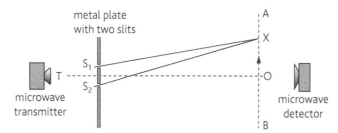

Figure 11.5.4 *Demonstrating interference of microwaves*

Key points

- An interference pattern of water waves can be produced in a ripple tank.
- Interference of sound waves can be produced using two loudspeakers connected to the same signal generator.
- Interference of microwaves can be produced by using a microwave source, a metal sheet with two slits and a microwave detector.

On completion of this section, you should be able to:

- derive the formula for the fringe separation in Young's double slit experiment

- describe an experiment to measure the wavelength of light using Young's double slit experiment

- describe an experiment to measure the wavelength of light using a diffraction grating.

Derivation of the formula $\lambda = \dfrac{ax}{D}$

Consider two coherent light sources S_1 and S_2 separated by a distance a. M is the midpoint of S_1 and S_2. The distance between the slits and the screen is D. A central bright fringe is located at the point O. In this case, the path difference is equal to zero. Constructive interference occurs at O. The first bright fringe occurs at the point P (Figure 11.6.1).

The formula can be derived by considering the distance between the central bright fringe and the first adjacent bright fringe. The path difference $S_2P - S_1P$ is equal to one wavelength λ.

Consider the triangle PMO.

$$\tan\theta = \frac{x}{D}$$

Consider the triangle S_2S_1Q

$$\sin\theta = \frac{\lambda}{a}$$

If $D >> a$, the angle θ is very small.

For small angles $\quad \theta \approx \sin\theta \approx \tan\theta$

$$\therefore \quad \frac{\lambda}{a} = \frac{x}{D}$$

$$\therefore \quad \lambda = \frac{ax}{D}$$

The equation is only applicable when $D \gg a$ and that S_1 and S_2 act as coherent sources of light.

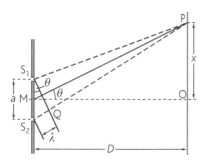

(diagram not drawn to scale)

Figure 11.6.1 *Deriving the formula for wavelength*

Measuring the wavelength of light using the Young's double slit arrangement

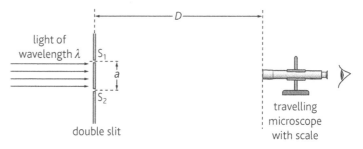

(diagram not drawn to scale)

Figure 11.6.2 *Measuring the wavelength of light using Young's double slit experiment*

The wavelength of a monochromatic light source can be determined using Young's double slit arrangement. In this experiment a parallel beam of monochromatic light of wavelength λ, is incident normally on a pair of slits S_1 and S_2. The width of the slits is approximately 0.5 mm and the distance between the slits is approximately 1 mm. In this experiment, a travelling microscope is used to measure the distance between the bright fringes produced (Figure 11.6.2). The distance between the slits and the travelling microscope is approximately 2 m. In order to measure the distance between

the fringes accurately, the distances between several fringes are measured and the average separation taken. This minimises the random error in the measurement. The disadvantage of this method is the intensity of the light decreases as you move away from the central bright fringe. This makes it difficult to know exactly where a bright fringe is located. The wavelength of the monochromatic light source is calculated as shown (right).

Measuring the wavelength of monochromatic light using a diffraction grating

A laser can be used as a monochromatic light source for this experiment. A laser is used because it produces a highly collimated light source. This means that it produces a very thin beam of light with little spreading. A diffraction grating of known number of lines per metre N is used. The diffraction grating is set up on a spectrometer. The light from the laser is projected on the diffraction grating such that it is at right angles to it (Figure 11.6.3). The angular deviation can be measured accurately using the spectrometer. The wavelength of the light is calculated as shown (right).

The experiment can also be done using a screen. The screen is placed about 1.5 m from the diffraction grating. The separation between the zero-order diffracted light and the nth-order diffracted light is measured using a metre rule. The angular deviation of the nth order diffracted light is found by calculation.

One advantage of measuring the wavelength using the second-order diffracted light is that it allows for a larger distance to be measured accurately.

One disadvantage of measuring the wavelength using the second-order diffracted light is that it is difficult to pinpoint its actual position, because intensity of the light decreases as you move away from the zero-order diffracted light.

Equation

$$\lambda = \frac{ax}{D}$$

x – fringe separation/m
λ – wavelength of light/m
D – perpendicular distance between the screen and the double slits/m
a – the separation between the slits/m

Equation

$$\lambda = \frac{d \sin\theta}{n}$$

λ – wavelength of monochromatic light/m
d – distance between slits/m
θ – angular deviation for the nth order
n – nth order diffracted light

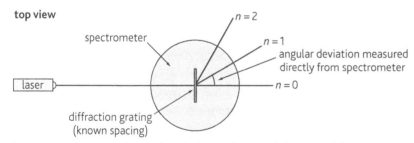

top view

spectrometer

laser

diffraction grating (known spacing)

$n = 2$
$n = 1$
angular deviation measured directly from spectrometer
$n = 0$

Figure 11.6.3 *Measuring the wavelength of monochromatic light using a diffraction grating*

Example

A laser produces monochromatic light which is incident at right angles to a diffraction grating. The diffraction grating has 5.6×10^5 lines per metre. A screen is located 1.50 m from the diffraction grating. Bright spots are observed at 0.54 m on either side of the central bright spot. Calculate the wavelength of the monochromatic light.

Distance between slits $d = \frac{1}{N} = \frac{1}{5.6 \times 10^5} = 1.79 \times 10^{-6}$ m

Angle between zero-order and first-order light $\theta = \tan^{-1}\left(\frac{0.54}{1.5}\right) = 19.8°$

$$\lambda = \frac{d \sin\theta}{n} = \frac{1.79 \times 10^{-6} \times \sin 19.8°}{1} = 606 \, \text{nm}$$

Key points

- The wavelength of a monochromatic light source can be determined experimentally using Young's double slit experiment.

- The wavelength of a monochromatic light source can be determined experimentally using a diffraction grating.

Learning outcomes

On completion of this section, you should be able to:

■ understand what is meant by a stationary wave

■ describe experiments to produce stationary waves

■ compare progressive and stationary waves.

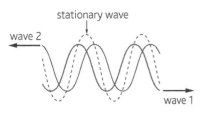

Figure 11.7.1 *Producing a stationary wave*

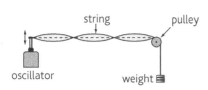

Figure 11.7.2 *Producing a stationary wave on a string*

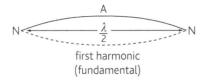

first harmonic
(fundamental)

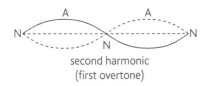

second harmonic
(first overtone)

third harmonic
(second overtone)

Figure 11.7.3

Stationary waves

Stationary waves (standing waves) are formed by the superposition of two progressive waves of the same type, of equal amplitude and frequency travelling with the same speed in opposite directions. Both longitudinal and transverse waves can form stationary waves. Figure 11.7.1 shows a transverse progressive wave travelling from left to right. Another transverse progressive wave, of same speed, frequency and amplitude is travelling from right to left. These two progressive waves superpose to produce a **stationary wave** (dashed line).

A stationary wave can be produced on a string by causing it to oscillate rapidly. In this experiment, one length of a long string is attached to a mechanical oscillator. The string passes over a frictionless pulley and is kept taut by means of several small weights. The mechanical oscillator produces progressive waves which travel from left to right. The waves are reflected as they reach the pulley and travel from right to left. The waves leaving the oscillator superpose with the reflected waves and produce a stationary wave. See Figure 11.7.2.

The velocity of the incident wave and the reflected wave along the string is given by:

Equation

$$v = \sqrt{\frac{T}{m}}$$

v – velocity of wave/$m\,s^{-1}$

T – tension in the string/N

m – mass per unit length of the string/$kg\,m^{-1}$

Suppose a stretched string is fixed at its end and then made to vibrate. Only specific modes of vibrations are possible on the string. Figure 11.7.3 illustrates the three simplest modes of vibrations (i.e. first, second and third harmonics). The first harmonic is called the fundamental and the frequency at which it vibrates is called the **fundamental frequency**. The higher frequencies are called overtones.

The fundamental frequency, f_0, of a vibrating, stretched string is given by:

Equation

$$f_0 = \frac{1}{2l}\sqrt{\frac{T}{m}}$$

f_0 – fundamental frequency/Hz

l – length of string/m

T – tension in the string/N

m – mass per unit length/$kg\,m^{-1}$

There are points along a stationary wave where particles are permanently at rest. These points are called **nodes**. The displacement at these points is zero. At points midway between nodes on a stationary wave the displacement is twice the amplitude of either progressive wave. These points are called **antinodes** (Figure 11.7.4).

The wavelength of the wave is given by:

Equation

$\lambda = 2d$

λ – wavelength of wave/m
d – distance between two adjacent nodes/m

Figure 11.7.5 shows the position of the string at various points in time. From this diagram it can be seen that all the particles between adjacent nodes are in phase with each other. All the particles between N_1 and N_2 are in phase with each other. All the particles between N_2 and N_3 are also in phase with each other. However, the particles between N_1 and N_2 are out of phase with the particles between N_2 and N_3. All the particles between N_1 and N_2 are moving downwards, while the particles between N_2 and N_3 are moving upwards.

Consider the phase of the particles in a progressive wave. Each point within one cycle of the wave is at a different phase from each other. Therefore, points A, B, C and D are at a different phase in their cycle. Points A and D are moving upwards, while points B and C are moving downwards.

Points A and E are in phase with each other. They are both moving upwards.

Points C and F are in phase with each other. They are both moving downwards (Figure 11.7.6).

The amplitude of the stationary wave depends on the position along the wave. It varies from zero at a node to a maximum at the antinode.

The speed of sound can be measured using a long cylindrical tube with one end open and the other closed as shown in Figure 11.7.7. The tube is lined with a fine powder. A loudspeaker, connected to a signal generator, is placed at the open end. The frequency of the signal generator is adjusted so that the powder eventually settles. The sounds waves inside the tube create a stationary wave. The powder settles at displacement nodes. The distance between the nodes is used to determine the wavelength of the sound waves and since the frequency is known, the speed of the wave can be determined.

Comparing stationary and progressive waves

Progressive wave	Standing wave
Transfers energy from one point to another.	Even though the wave has energy, it does not transfer energy from one point to another.
All the particles that make up the wave have the same amplitude.	The particles that make up the wave have different amplitudes. The amplitude ranges from zero (nodes) to a maximum (antinodes).
All the particles that make up the wave are in motion.	There are particles that make up the wave that are stationary (antinodes).
A phase difference exists between neighbouring particles of the wave.	Between two nodes the phase difference between neighbouring particles is zero. (A and B)
	The phase difference between particles A and C is π radians.

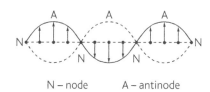

N – node   A – antinode

Figure 11.7.4 *Nodes and antinodes*

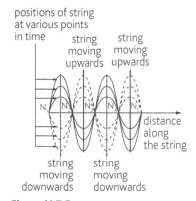

Figure 11.7.5

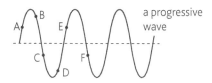

Figure 11.7.6 *A progressive wave*

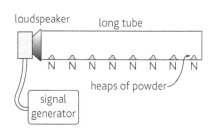

Figure 11.7.7 *Producing a stationary sound wave*

Key points

- Stationary waves (standing waves) are formed by the superposition of two progressive waves of the same type, of equal amplitude and frequency travelling with the same speed in opposite directions.

- A node is a point on a wave that is permanently at rest.

- An antinode is the point midway between nodes.

11.8 Sound waves

Learning outcomes

On completion of this section, you should be able to:

- describe practical applications of sound waves in industry

- discuss the application of sound waves to musical instruments

- understand that sound waves can be reflected and refracted.

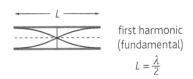

first harmonic (fundamental)

$$L = \frac{\lambda}{2}$$

second harmonic (first overtone)

$$L = \lambda$$

third harmonic (second overtone)

$$L = \frac{3\lambda}{2}$$

Figure 11.8.1 *Stationary waves in an open-ended tube*

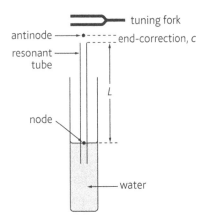

Figure 11.8.2 *A resonance tube*

Sound waves in industry

Sound waves are used widely in industry. In the field of medicine, ultrasound (very high frequency sound waves) is used to produce images of foetuses in pregnant women, and internal organs such as the kidney and liver. Ultrasound is also used in physiotherapy.

The same principle is used to determine if there are hairline fractures in mechanical components used in industries.

Submarines use sonar to determine depths of objects. It operates in a similar way to how bats determine the location of objects. Sound waves are transmitted from the submarines and the time taken for the reflected wave is determined. The distances can then be calculated.

Geologists use sound waves to determine the structure of the underlying earth. In countries such as Trinidad this technique is crucial in determining the location of oil and gas.

Sound waves and musical instruments

The lowest frequency that a vibrating string or pipe can produce is called the fundamental frequency. This note is called the **fundamental**.

If a note is n times the fundamental, it is called the nth **harmonic**.

When a string of a guitar is plucked, notes of high frequencies are produced together with the fundamental note. These notes are called **overtones**.

A musical note is characterised by **loudness, pitch and timbre**. Loudness of a note depends on the amplitude of sound. Pitch of a note is dependent on the frequency of sound. Timbre or quality of a note is dependent on the relative strengths of the overtones produced with the note.

Percussion instrument – steel pan

The steel pan is a percussion instrument. It produces sound when struck by a stick with a piece of rubber attached to its end. When a particular note is struck, stationary waves are set up on that particular section of the steel pan. Only certain modes of vibration are possible. The largest amplitude of vibration is that of the fundamental frequency. Overtones are also present and give the characteristic 'steel pan' sound.

Stringed instrument – guitar

A guitar has several strings attached to it. When a string is plucked, transverse waves travel along the string and are reflected on reaching the ends. Stationary waves are produced (refer to Figure 11.7.3). Only certain modes of vibrations are allowed. The string vibrates with the fundamental frequency as well as the overtones.

Wind instrument – flute

A flute is made of a hollow tube. Both ends are open. When air is blown across one end, the air inside the tube begins to vibrate. The vibration produces a longitudinal wave (pressure wave) which travels along the tube and is reflected at the far end. The incident longitudinal wave and the reflected wave superpose and a stationary wave is produced. As in the case of the steel pan and guitar, only certain modes of vibrations are allowed.

Resonance tubes

A resonance tube can be used to determine the speed of sound. Figure 11.8.2 shows the arrangement of the apparatus used to perform the experiment. A tuning fork of known frequency is placed over the air column. The air above the tube vibrates. A longitudinal wave is transmitted down the tube and is reflected when it hits the surface of the water. The two waves superpose and produce a stationary wave. Depending on the length of the air column, only certain modes of vibrations are possible. If the fundamental mode of vibration corresponds to the frequency of the tuning fork, resonance occurs.

At resonance, the sound heard inside the tube is enhanced (a loud sound is heard). Resonance will occur at specific lengths when the length of the air column is slowly increased. These lengths will correspond to the different modes of vibrations possible. A node will exist at the surface of the water. An antinode will occur just above the open end of the resonance tube. An end-correction of length c is sometimes used in calculations to cater for this (Figure 11.8.3).

Example

A small loudspeaker emitting sound of constant frequency is positioned a short distance above a long glass tube containing water. When water is allowed to run slowly out of the tube, the intensity of the sound heard increases at specific lengths of air in the tube. When the loudspeaker is emitting sound at a frequency of 400 Hz, the effect is first noticed at length = 21 cm. It next occurs at length = 64 cm.

a State the phenomenon taking place.

b Calculate the wavelength of the sound waves in the air column.

c Calculate the speed of the sound waves in the air column.

a Resonance is taking place in the tube.

b When resonance first occurs, $\frac{\lambda}{4} = 21$, $\therefore \lambda = 4 \times 21 = 84$ cm

When resonance occurs again, $\frac{3\lambda}{4} = 64$, $\therefore \lambda = \frac{4 \times 64}{3} = 85.3$ cm

c Average wavelength $= \frac{84 + 85.3}{2} = 84.7$ cm

Speed of sound $= f \times \lambda = 400 \times 84.7 \times 10^{-2} = 339$ m s^{-1}

Reflection and refraction of sound waves

Sound waves can be reflected as echoes. Acoustical engineers design concert halls so that the amount of reflection of sound waves is limited. Sound absorbing materials are used to line the walls.

Refraction occurs when the speed of a wave changes. Sound waves require a material medium in order to be transmitted. The speed of sound waves in air depends on the temperature of the air. A sound wave travelling through layers of air of different temperatures will be refracted. On a sunny day, the ground heats up the lower layers of the atmosphere. The upper layers are cooler than the lower layers. When someone shouts, sound waves are refracted upwards away from the ground. At night time, the temperature of the lower layers of the atmosphere is lower than the upper layers. When someone shouts, sound waves are refracted downwards towards the ground.

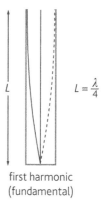

first harmonic
(fundamental)

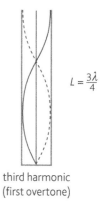

third harmonic
(first overtone)

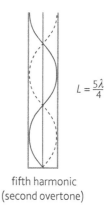

fifth harmonic
(second overtone)

Figure 11.8.3 *Modes of vibration in a resonance tube*

Key points

- Sound waves have many applications in industry.

- Sound waves are produced by various instruments.

- Sound waves can be reflected and refracted.

Properties of electromagnetic waves

An electromagnetic wave consists of an oscillating electric field and magnetic field at right angles to each other. Figure 11.9.1 shows a graphical representation of such a wave.

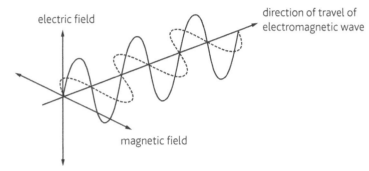

Figure 11.9.1 An electromagnetic wave

Properties of electromagnetic waves:

- They travel at the speed of light $(3 \times 10^8\,\mathrm{m\,s^{-1}})$ in a vacuum.
- They consist of oscillating electric and magnetic fields at right angles to each other.
- They are transverse waves.
- They can travel in a vacuum.
- They can be refracted, diffracted, reflected and polarised.

The electromagnetic spectrum

Electromagnetic waves can be arranged based on the magnitude of their wavelengths. Radio waves have the largest wavelength, while gamma rays have the shortest wavelength. They form part of what is referred to as the **electromagnetic spectrum** (Table 11.9.1).

Table 11.9.1 The electromagnetic spectrum

Electromagnetic wave	Frequency/Hz	Wavelength/m
Radio waves	Lowest frequency	$>10^{-1}$
Microwaves		10^{-2} to 10^{-4}
Infrared		$\sim 10^{-6}$
Visible		7×10^{-7} to 4×10^{-7} (ROYBGIV) (Red ⟶ Violet)
Ultraviolet		$\sim 10^{-8}$
X-rays		10^{-9} to 10^{-11}
Gamma rays	Highest frequency	10^{-12} to 10^{-16}

In the visible part of the electromagnetic spectrum, red light has the longest wavelength (640 nm) and violet light (470 nm) has the shortest wavelength.

Table 11.9.2 *Sources of em waves and their uses*

Type of electromagnetic wave	Source	Use
Radio waves	▪ Transmitters	▪ Telecommunications ▪ Navigation systems
Microwaves	▪ Klystron tubes ▪ Microwave ovens	▪ Telecommunications ▪ Heating of food ▪ Radar
Infrared	▪ All hot objects	▪ Enables pictures to be taken in the dark. ▪ Used in fibre optic cables – telecommunications ▪ Remote control for your TV
Visible	▪ Hot objects ▪ The Sun ▪ Incandescent and fluorescent lamps	▪ To be able to see objects.
Ultraviolet	▪ Mercury vapour lamp ▪ Electric arcs (sparks)	▪ To kill bacteria ▪ Produces Vitamin D in the skin
X-rays	▪ X-ray tubes	▪ Medical imaging – Radiography ▪ Cancer treatment ▪ Used in industry to check the quality of welded joints and detect 'hair line' fractures in metals. ▪ To determine the structure of crystals
Gamma rays	▪ Cosmic radiation ▪ Nuclear power plants ▪ Radioactive elements	▪ Cancer treatment ▪ Sterilisation of surgical instruments

Key points

- Electromagnetic waves consist of oscillating electric and magnetic fields at right angles to each other.
- Electromagnetic waves have similar properties (e.g. same speed in a vacuum, transverse, can be diffracted, can be polarised)
- The electromagnetic spectrum consists of a list of electromagnetic waves arranged in order of increasing wavelengths.

Revision questions 5

Answers to questions that require calculation can be found on the accompanying CD.

1 Explain, using clearly labelled diagrams where appropriate, the meaning of the following terms:
   a displacement [1]
   b amplitude [1]
   c phase of a particle in a wave [1]
   d the phase difference between two waves [2]
   e coherent waves [2]

2 a Explain what is meant by the following terms when applied to waves:
   i wavelength [1]
   ii frequency [2]
   iii speed [1]
   b Using these definitions, deduce an equation for the speed of a wave. [3]

3 a Distinguish between transverse and longitudinal waves. [2]
   b State one example of each type of wave. [2]
   c State two similarities between the two types of waves. [2]
   d State and describe with the aid of a diagram, one wave phenomenon that applies to transverse waves only. [3]

4 a Explain what is meant by plane polarisation of light waves. [1]
   b Explain why sound waves cannot be plane-polarised. [2]

5 a Explain what is meant by the term *diffraction*. [2]
   b Describe how the width of the gap through which a wave is passing affects the diffraction of a wave. (Use a diagram to illustrate both a narrow gap and a wide gap.) [3]
   c A beam of blue light from a monochromatic light source is projected at right angles to a diffraction grating. Bright light is seen emerging from the diffraction grating at certain angles. Use the principle of superposition to explain this effect. [2]

6 Explain what is meant by the principle of superposition. [2]

7 When monochromatic light of wavelength λ is incident normally on a plane diffraction grating, the second order diffraction lines are formed at angles of 28° to the normal to the grating. The diffraction grating has 600 lines per millimetre. Calculate the value of λ. [4]

8 Determine the highest order of diffracted beams that can be produced when a grating with a spacing of 2.0×10^{-6} m is illuminated normally with light of wavelength 400 nm. [3]

9 In order that interference of waves from two sources may be observed, the waves must be coherent.
   Explain what is meant by:
   i coherence   ii interference. [4]

10 State three conditions necessary to observe the interference of light. [3]

11 a Describe an experiment to demonstrate the interference of microwaves. [5]
   b Using the principle of superposition, explain the results of this experiment. [4]

12 a Describe an experiment to demonstrate interference in a ripple tank. [4]
   b Explain how coherent waves are produced in the ripple tank. [2]
   c Draw a diagram to show the interference pattern produced on the water. Label two points with the letter C, where constructive interference occurs. Label two points with the letter D, where destructive interference occurs. [4]

13 An interference pattern is produced on a screen when coherent red light passes through a double-slit arrangement. State with a reason, the effect on the appearance and spacing of the fringes observed when, independently, the following changes are made:
   a Coherent blue light is used as the source instead of red light. [2]
   b The distance between the slits is decreased. [2]
   c The width of each slit in the double-slit arrangement is gradually increased. [3]
   d One of the slits is covered by an opaque card. [1]

14 Young's double slit experiment can be used to measure the wavelength of monochromatic light.
   a Describe Young's double slit experiment. [6]
   b State what measurements are taken and explain how these measurements are used to calculate the wavelength. [4]
   c State approximate values for:
   i the width of each slit [1]
   ii the distance between the two slits [1]
   iii the distance between the double slit and the screen. [1]

d Explain the roles played by diffraction and interference in the production of the observed fringes. [4]

15 A laser produces monochromatic red light of wavelength 620 nm which is incident at right angles to a diffraction grating. The diffraction grating has 5.5×10^5 lines per metre. A screen is located 1.50 m from the diffraction grating. Bright spots are observed at different points on the screen. Calculate the distance between the central bright spot and the second-order diffracted light. [4]

16 Distinguish between a stationary wave and a progressive wave by making reference to:
   a the amplitude of vibration of particles in each wave [2]
   b the phase difference between neighbouring particles in each wave [3]
   c the energy transferred along each wave. [2]

17 A string of length 80 cm is stretched between two fixed points. It is plucked at its centre and the string vibrates to form a stationary wave. A single antinode is formed at the centre of the string.
   a Explain:
      i what is meant by a stationary wave [2]
      ii what is meant by an antinode [2]
      iii why nodes are formed at the ends of the string. [1]
   b State the wavelength of the wave. [1]
   c The frequency of vibration of the string is 340 Hz. Calculate the speed of the wave. [2]
   d Explain what is meant by the speed calculated in c. [2]

18 One end of a long string is attached to an oscillator. The string passes over a frictionless pulley and is kept taut by means of a weight of 3.50 N. The frequency of the oscillator is adjusted so that a stationary wave is produced. Three antinodes are present on the wave. The distance between the two adjacent antinodes is 16.2 cm and the frequency is 118 Hz. Calculate the mass per unit length of the string. [3]

19 State three subjective sensations to sound and state what they depend on. [6]

20 A small loudspeaker emitting sound of constant frequency is positioned a short distance above a long glass tube containing water. When water is allowed to run slowly out of the tube, the intensity of the sound heard increases at specific lengths of air in the tube. When the loudspeaker is emitting sound at a frequency of 400 Hz, the effect is first noticed at length = 19.7 cm. It next occurs at length = 59.1 cm.
   a Calculate the wavelength of the sound waves in the air column. [3]
   b Calculate the speed of the sound waves in the air column. [3]

21 a State three features of waves which are common to all regions of the electromagnetic spectrum. [3]
   b State a typical wavelength of:
      i red light [1]
      ii ultraviolet radiation [1]
      iii infrared radiation [1]
      iv X-rays. [1]

22 Arrange the following in increasing magnitude of wavelength : X-rays, microwaves, ultraviolet waves and infrared waves. [4]

12.1 The physics of hearing

12.1 The physics of hearing

Learning outcomes

On completion of this section, you should be able to:

- explain how the ear responds to an incoming sound wave
- understand the significance of the terms *sensitivity* and *frequency response*
- state the magnitude of the threshold of hearing
- state the intensity at which discomfort is experienced
- use the equation for intensity level
- understand the terms *noise* and *loudness*.

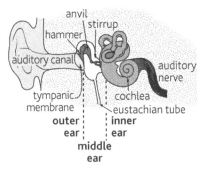

Figure 12.1.1 *The human ear*

Equation

Intensity is the power per unit area

$I = \dfrac{P}{A}$ at a stated frequency.

I – intensity/W m^{-2}
P – power/W
A – area/m^2

The human ear

Figure 12.1.1 shows a diagram of the human ear.

The outer ear collects and directs sound waves into the auditory canal. At the end of the auditory canal is the tympanic membrane (ear drum). The sound waves cause the tympanic membrane to vibrate. The middle ear consists of three tiny bones called the hammer, the anvil and the stirrup. When the tympanic membrane vibrates, the bones in the middle ear vibrate as well. The three bones act like a lever system. They reduce the amplitude of vibration produced on the tympanic membrane. At the same time, the vibrational pressure on the oval window is increased.

The middle ear is connected to back of the throat via the eustachian tube. Under normal conditions, the pressure on both sides of the tympanic membrane is the same. When someone ascends in an aircraft, pressure difference exists on either side of the tympanic membrane resulting in the ear "popping".

The inner ear is filled with liquid. It contains the cochlea, which is attached to auditory nerve. Inside the cochlea there are many tiny hairs that run its entire length. These hairs vary in length, thickness and stiffness. When the oval window vibrates, the liquid inside the inner ear vibrates causing the tiny hairs to resonate. The hairs produce electrical signals which are then transmitted via the auditory nerve to the brain, where they are then interpreted.

Frequency response and intensity

The human ear is able to detect frequencies in the range 20 Hz to 20 kHz. This range is called the **frequency response** of the ear. As a person gets older, the upper limit of 20 kHz decreases. In the frequency range 60 Hz to 1 kHz, the human ear can detect changes of 2 Hz to 3 Hz. At frequencies above 1 kHz, it is difficult for the human ear to detect small changes in frequencies.

Intensity is the sound power per unit area $\left(I = \dfrac{P}{A}\right)$ at a stated frequency.

The smallest sound intensity that can be detected by the human ear is called the **threshold of hearing**. The threshold of hearing is 1.0×10^{-12} W m^{-2} at a frequency of 3 kHz. The threshold of hearing varies with frequency.

Figure 12.1.2 shows variation of the threshold of hearing with frequency of a certain person. Above the curve represents intensities that can be detected by the human ear. Intensities below the curve cannot be detected by the human ear. The **sensitivity** of the human ear is the ability to detect the smallest fractional change ΔI of intensity I. Sensitivity depends on the ratio $\dfrac{\Delta I}{I}$.

Sensitivity increases with frequency to a maximum and then decreases with increasing frequency. The maximum sensitivity is at 1–3 kHz.

The human ear can detect a wide range of intensities. The minimum intensity that can be detected is 1.0×10^{-12} W m^{-2} at a frequency of 3 kHz. The upper limit of the range is 100 W m^{-2} in the frequency range 1 kHz to 6 kHz. The upper limit is called the **threshold of pain**. Persons exposed to intensities of 100 W m^{-2} can experience pain and temporary deafness.

The logarithmic response of the ear

Equal changes in intensity are not perceived as equal changes in loudness.

Loudness is the subjective response of a person to a given intensity. Intensity level may be used as a measure of loudness. **Intensity level** is defined using the following equation. Changes in loudness depend on the fractional change in intensity $\left(\dfrac{\Delta I}{I}\right)$. Loudness is a logarithmic response to intensity.

Example

A person with normal hearing is exposed to a sound of frequency 3 kHz and an intensity level at the ear of 20 dB. Calculate the intensity of this sound at the ear.

$$\text{Intensity level} = 10 \log_{10}\left(\frac{I}{I_0}\right)$$

$$20 = 10 \log_{10}\left(\frac{I}{1 \times 10^{-12}}\right)$$

$$\log_{10}\left(\frac{I}{1 \times 10^{-12}}\right) = 2$$

$$\frac{I}{1 \times 10^{-12}} = 100$$

$$I = 1 \times 10^{-10}\,\text{W m}^{-2}$$

Loudness and noise

Intensity level is defined in terms of the intensity of sound and the threshold of hearing at a frequency of 3 kHz.

Loudness is the subjective response of a person to a given intensity and depends on the individual and the frequency of the sound.

To define a consistent unit of loudness, a frequency of 1 kHz is chosen as a standard. The 1 kHz standard source is adjusted until it is perceived as being as loud as the source being evaluated. Suppose the intensity level of the 1 kHz standard source is found to be 90 dB when it is as loud as the test source, then the loudness of the test source is 90 phons.

Example

The intensity of sound measured at a distance of 0.8 m from a loudspeaker is measured as $1.2 \times 10^2\,\text{W m}^{-2}$. Given that the intensity of sound varies as $\dfrac{1}{d^2}$, where d is the distance measured from the loudspeaker, calculate the intensity of the sound at a distance of 3.0 m.

$$I \propto \frac{1}{d^2}$$

$$I = \frac{k}{d^2}$$

$$1.2 \times 10^2 = \frac{k}{(0.8)^2}$$

$$k = 76.8$$

Therefore the intensity at a distance of 3.0 is given by:

$$I = \frac{k}{d^2} = \frac{76.8}{(3.0)^2} = 8.53\,\text{W m}^{-2}$$

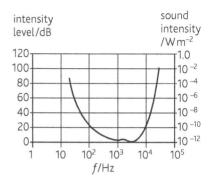

Figure 12.1.2 *Variation of threshold of hearing with frequency*

Equation

$$\text{Intensity level} = 10 \log_{10}\left(\frac{I}{I_0}\right)$$

Intensity level is measured in dB (decibels)

I – intensity of sound incident on ear/W m^{-2}

I_0 – threshold of hearing ($1.0 \times 10^{-12}\,\text{W m}^2$)

Key points

- The human ear is able to detect frequencies in the range 20 Hz to 20 kHz.

- The threshold of hearing is the minimum intensity that can be detected by the ear.

- The sensitivity of the human ear is the ability to detect the smallest fractional change ΔI of intensity I.

- Loudness is the subjective response of a person to a given intensity.

- The ear has a logarithmic response to sound.

13.1 Lenses

Lenses

A converging or convex lens is thicker at the middle than at the edges. When a parallel beam of light strikes the surface of a converging lens, refraction occurs and the beam is focused through a point known as the focal point F of the lens (Figure 13.1.1). The point P is known as the optical centre or pole P of the lens. The horizontal line drawn through the optical centre of the lens is known as the **principal axis**. The distance *f* between the optical centre of the lens P and the focal point F is known as the **focal length** of the lens.

The rays of light actually pass through the point F. If a screen is placed at the point F, an image will be seen. The image produced in this case is called a **real image**.

A diverging or concave lens is thinner at the centre than at the edges. When a parallel beam of light strikes the surface of a diverging lens, refraction occurs and the rays spread out as they leave the lens. The rays appear to have diverged from the point F. If a screen is placed at the point F, an image will not be seen, because the rays do not actually pass through the point F. The image in this case is called a **virtual image**. The distance between the optical centre of the lens and the point F is called the focal point of the diverging lens.

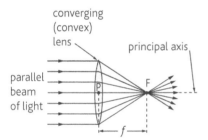

Figure 13.1.1 *A converging (convex) lens*

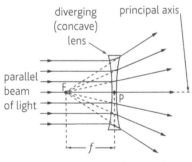

Figure 13.1.2 *A diverging (concave) lens*

The power of a lens is defined by the equation here.

The unit for power of a lens is the dioptre. Its symbol is D.

A more powerful lens will have a shorter focal length (Figure 13.1.3).

Converging lenses have positive values.

Diverging lenses have negative values.

For example:

- A converging lens of focal length 0.4 m will have a power of

$$+\frac{1}{0.4} = +2.5\,\text{D}.$$

- A diverging lens of focal length −0.3 m will have a power of

$$-\frac{1}{0.3} = -3.33\,\text{D}.$$

Equation

$$P = \frac{1}{f}$$

P – power of a lens/dioptres
f – focal length of lens/m

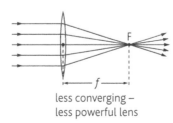

less converging –
less powerful lens

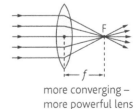

more converging –
more powerful lens

Figure 13.1.3 *Relating power and focal length of a lens*

The lens formula

An object is placed in front (to the left) of a converging lens. The distance between the object and optical centre of the lens is the object distance u. To determine the position of the image several rays are constructed.

Rules for constructing ray diagrams:

- A ray of light parallel to the principal axis passes through the focal point of the lens.
- A ray of light passing through the optical centre of the lens is undeviated.

Using these rules, an image can be constructed on the right-hand side of the lens. Note that the image is inverted and is real. The distance between the image and the optical centre of the lens is the image distance v.

The lens formula (shown right) can be used for converging and diverging lenses. A sign convention is used (real is positive).

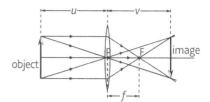

Figure 13.1.4

	Quantity	Positive sign (+) (Real)	Negative sign (−) (Virtual)
1	Object distance, u	Object is in front of lens	Object is at the back of lens
2	Image distance, v	Image is at the back of lens	Image is in front of lens
3	Focal length, f	Converging (convex) lens	Diverging (concave) lens

- The focal length of a converging lens is positive.
- The focal length of a diverging lens is negative.
- Real object and real image distances are positive.
- Virtual object and virtual image distances are negative.

Example

An object is placed 12 cm in front of a converging lens of focal length 20 cm. Calculate the position of the image and state its nature.

$u = +12$ cm (real object) $f = +20$ cm (converging lens)

$$\frac{1}{f} = \frac{1}{u} + \frac{1}{v}$$

$$\frac{1}{20} = \frac{1}{12} + \frac{1}{v}$$

$$\frac{1}{v} = \frac{1}{20} - \frac{1}{12} = -\frac{1}{30}$$

$$v = -30 \text{ cm}$$

Since the image distance is negative, it is virtual and is in front of the converging lens (same side of the lens as the object).

A simple camera

A simple camera is shown in Figure 13.1.5. The lens of the camera can move back and forth from the photograph film. This allows for light from objects at different distances to be focused on the film. The image produced is real, diminished and inverted.

A magnifying glass

When an object is placed between the focal point and the lens, a virtual, upright and enlarged image is produced. This is the principle by which a magnifying glass works.

Equation

$$\frac{1}{f} = \frac{1}{u} + \frac{1}{v}$$

f – focal length of the lens/m
u – object distance/m
v – image distance/m

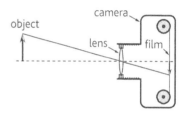

Figure 13.1.5 A simple camera

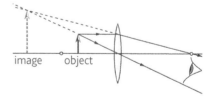

Figure 13.1.6 A magnifying glass

Key points

- A converging lens focuses a beam of parallel rays through a single point called the focal point.
- A diverging lens spreads a beam of parallel rays so that they appear to have diverged from a single point called the focal point.

The human eye

Figure 13.2.1 illustrates the human eye. The rays of light entering the eye must pass through several media (cornea, aqueous humour, lens, vitreous humour) before reaching the retina. Each medium has a different refractive index. Light first strikes the air-cornea boundary. Most of the bending occurs at this boundary because of the large difference between the refractive indices (refractive index of air = 1.0, refractive index of cornea = 1.38). The light then travels through the aqueous humour, then through the pupil towards the lens. The main function of the lens is to fine tune the focusing of light so that an image is formed on the retina. The retina consists of nerve endings that generate electrical impulses that are sent to the brain via the optic nerve.

The lens is suspended by ligaments which are attached to a circular ring of muscles called the ciliary muscles. When the muscles are relaxed, the lens is long and thin. When the muscles contract, the lens becomes short and fat (more powerful or shorter focal length).

Accommodation is the ability of the eye to change the focal length of the lens so as to focus images formed from objects at different distances.

The closest point to the eye at which the eye can still produce a focused image on the retina is 25 cm for a normal eye and is called the near point. The far point of the normal eye is taken to be at infinity.

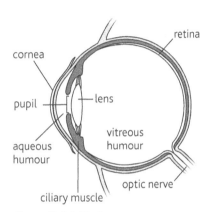

Figure 13.2.1 *The human eye*

Depth of field and depth of focus

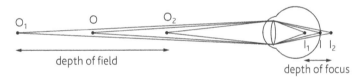

Figure 13.2.2 *Depth of field and depth of focus*

Consider an object O being viewed at some distance from eye. The depth of field of the eye is the distance moved by the object (between O_1 and O_2), while the image remains in focus (Figure 13.2.2).

For a given accommodation, the eye is able to see clearly an object slightly nearer (O_2) and slightly further (O_1) from some fixed point. This variation in distance through which the eye can still see clearly is called the depth of focus.

The depth of field and the depth of focus are affected by the size of the iris.

Short-sightedness (myopia)

The person is able to only focus objects close to the eye. Distant objects are blurred. The image forms in front of the retina. Short-sightedness occurs when the lens is not able to relax in order to become long and thin (less powerful). Short-sightedness also occurs if the eyeball is too long. A diverging lens is used to correct short-sightedness. In the case of short-sightedness, the far point is closer than infinity and the near point may be closer than 25 cm (Figures 13.2.3 and 13.2.4).

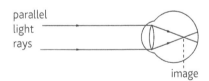

Figure 13.2.3 *Short-sightedness*

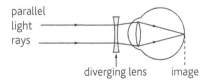

Figure 13.2.4 *Correction for short-sightedness*

Long-sightedness (hypermetropia)

The person is able to only focus objects that are far from the eye. Objects close to the eye are blurred. The image forms behind the retina. Long-sightedness occurs because the ciliary muscles become weak and the lens is not able to become short and fat (more powerful). Long-sightedness also occurs if the eyeball is too short. A converging lens is used to correct long sight. In the case of long-sightedness, the far point is infinity and the near point is greater than 25 cm (Figures 13.2.5 and 13.2.6).

Astigmatism

The person has difficulty focusing light rays from objects in different planes at the same time. The problem is caused because the surface of the cornea is uneven. Astigmatism is corrected using a cylindrical lens, adjusted such that its axis is perpendicular to the axis in which the eye cornea-lens system is cylindrical (Figure 13.2.7).

Cataracts

Persons with cataract have lenses which have become opaque. Very little light enters the eye. In order to correct this defect, the lens of the eye is removed. Surgeons can implant a new lens inside the eyes or glasses with converging lenses are used to correct the defect.

Example

A person has a near point of 30 cm and a far point of 90 cm. The cornea–retina distance is 1.7 cm. Calculate the power of:

a the eye when an object is viewed at the near point of 30 cm.

b the corrective lens needed to view an object at a distance of 25 cm.

c the corrective lens needed to view an object at infinity.

This person is short-sighted because the near point is greater than 25 cm.

The person is also long-sighted because the far point is less than infinity.

a $p = \dfrac{1}{f} = \dfrac{1}{u} + \dfrac{1}{v} = \dfrac{1}{0.30} + \dfrac{1}{0.017} = 62.2\,\text{D}$

b Long sight correction

$p = \dfrac{1}{f} = \dfrac{1}{u} + \dfrac{1}{v} = \dfrac{1}{0.25} + \dfrac{1}{0.017} = 62.8\,\text{D}$

Power of corrective lens $= 62.8 - 62.2 = +0.6\,\text{D}$ (Converging lens)

Alternatively, since the image at 30 cm is on the same side of the lens as the object, it is virtual. Therefore, the image distance is negative.

$p = \dfrac{1}{f} = \dfrac{1}{u} + \dfrac{1}{v} = \dfrac{1}{0.25} - \dfrac{1}{0.30} = 0.67\,\text{D}$

c Short sight correction

$p = \dfrac{1}{f} = \dfrac{1}{0.017} = 58.8\,\text{D}$

$p = \dfrac{1}{f} = \dfrac{1}{u} + \dfrac{1}{v} = \dfrac{1}{0.90} + \dfrac{1}{0.017} = 59.9\,\text{D}$

Power of corrective lens $= 58.8 - 59.9 = -1.1\,\text{D}$

Alternatively, the image distance is negative, because the image is virtual.

$p = \dfrac{1}{f} = \dfrac{1}{u} + \dfrac{1}{v} = \dfrac{1}{\infty} - \dfrac{1}{0.90} = -1.1\,\text{D}$

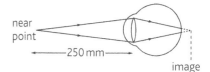

Figure 13.2.5 *Long-sightedness*

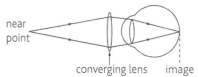

Figure 13.2.6 *Correction for long-sightedness*

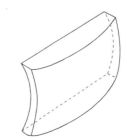

Figure 13.2.7 *Correction lens for astigmatism*

Key points

- Accommodation is the ability of the eye to focus images formed from objects at different distances.

- Depth of field is the distance moved by an object, while the image remains in focus.

- Common defects of the eye include short-sightedness, long-sightedness, astigmatism and cataracts.

- There are various techniques used to correct these defects.

Revision questions 6

Answers to questions that require calculation can be found on the accompanying CD.

1 a Explain what is meant by loudness and intensity level. [4]
  b Explain what is meant by the threshold of hearing and state its value. [3]
  c Describe how the sensitivity of the ear to loudness varies across the range of audible frequencies. [2]

2 a Explain what is meant by the 'frequency response' to sound waves of an average person. [3]
  b Describe and explain how this frequency response varies with age. [3]

3 A person has an eardrum of area 53 mm². When listening to music using headphones, the headphone produces 0.14 μW of sound power to the eardrum. Calculate:
  a the intensity of the sound incident on the eardrum [2]
  b the intensity level at the eardrum. [3]

4 a Describe how the ear responds to an incoming sound wave. [4]
  b Sketch a graph to show the variation with frequency of the intensity of sound at the ear for the threshold of normal hearing in a 15-year-old boy. [3]
    i Indicate, using values on the x-axis, the audible range of frequencies. [2]
    ii Indicate, using a value on the y-axis, the threshold of hearing. [1]
    iii State two ways in which the graph may be different for a 70-year-old man. [2]

5 a With the aid of a diagram, explain what is meant by the following terms:
    i focal length
    ii principal axis
    iii focal point. [3]
  b State the lens formula and write down an expression for the power of a lens. [3]

6 A lens has a power of +2.5 D. An object is placed 20 cm in front of the lens. Determine:
  a the focal length of the lens and state what type of lens it is [2]
  b the position and nature of the image produced. [3]

7 A diverging lens has a focal length of 20 cm. An object is placed at a distance of 30 cm from the optical centre of the lens. Calculate the position and nature of the image produced. [3]

8 State three properties of the image produced by a simple camera. [3]

9 Draw a ray diagram to explain how a magnifying glass works. [3]

10 Explain what is meant by:
  a accommodation of the eye [2]
  b depth of focus. [2]

11 The range of vision of an elderly gentleman is 1.1 m to 3.8 m. In order to correct his vision, an optometrist provides him with two pairs of spectacles. Calculate the power of the lens in each pair of spectacle. [3]

12 The power of the lens required to correct an eye defect is +1.5 D.
  a State the type of lens being used to correct the defect and calculate the focal length in cm. [2]
  b State the eye defect being corrected. [1]
  c Calculate the near point and far point of the unaided eye. [3]

13 a Explain what is meant by short-sightedness. [2]
  b Draw a diagram to show how this eye defect is corrected. [3]

14 a Explain what is meant by astigmatism. [2]
  b State the type of lens used to correct this eye defect. [1]

15 a Explain what is meant by the following terms:
    i accommodation [2]
    ii depth of field [2]
    iii depth of focus. [2]
b A student complains that she cannot see objects clearly unless they are more than 60 cm from her eyes.
    i State the student's eye defect. [1]
    ii State the near point of a normal person. [1]
    iii Calculate the power of the lens used to correct her defect. [3]
    iv Draw a ray diagram to show how two rays of light from a point object placed at the normal near point enters the eye. [3]
    v Draw another diagram to show how the correcting lens fixes her problem. [3]

16 a Explain what is meant by the following defects of the eye and state how they are corrected.
    i astigmatism [3]
    ii cataract. [2]

b An elderly person can see clearly objects which are situated between 1.2 m and 9 m from the eyes. Calculate the power of the lens required for him to see objects which are:
    i located at the near point of the eye [3]
    ii very distant from the eye. [3]

17 a Explain how the ear responds to an incoming sound wave. [4]
b A conversation is taking place between two students. The average intensity of the sound is $6.2 \times 10^{-6}\,\text{W m}^{-2}$.
    i Explain what is meant by the threshold of hearing and state is value. [2]
    ii Calculate the intensity level of the conversation. [3]
c Explain why the ear is said to have a logarithmic response to intensity. [3]

Module 2 Practice exam questions

Answers to the multiple-choice questions and to selected structured questions can be found on the accompanying CD.

Multiple-choice questions

1 A mass *m* is attached to a vertical helical spring having a spring constant *k*. The mass is displaced slightly and undergoes simple harmonic motion with amplitude *a*. The maximum velocity *v* of the mass is given by:

a $v = a\sqrt{\dfrac{m}{k}}$ b $v = a\sqrt{\dfrac{k}{m}}$

c $v = a\dfrac{m}{k}$ d $v = a\dfrac{k}{m}$

2 A particle P is undergoing simple harmonic motion. The displacement *x* in mm of the particle is given by $x = 0.58\cos(0.12\pi)t$. What is the frequency of oscillation?

a 0.06 Hz   b 0.58 Hz   c 0.12 Hz   d 0.22 Hz

3 A sound wave of amplitude 0.25 mm has and intensity of 3.5 W m⁻². What is the intensity of a sound wave of the same frequency which has an amplitude of 0.50 mm?

a 0.43 W m⁻²   b 14 W m⁻²
c 7.0 W m⁻²   d 1.75 W m⁻²

4 A stationary wave is produced when two similar progressive waves of frequency 400 Hz travelling in opposite directions superimpose. The distance between two adjacent nodes is 1.2 m. The speed of the progressive waves is:

a 480 m s⁻¹   b 900 m s⁻¹
c 960 m s⁻¹   d 167 m s⁻¹

5 A diffraction grating has *N* lines per metre. Monochromatic light of wavelength λ is incident at right angles to the grating. The angular deviation of the second order maxima is θ. Which of the following is true?

a $\sin\theta = 2N\lambda$ b $\sin\theta = 2\dfrac{\lambda}{N}$

c $\sin\theta = N\dfrac{\lambda}{2}$ d $\sin\theta = \dfrac{1}{2N\lambda}$

6 Ultraviolet rays differ from microwaves, in that ultraviolet rays:

a have a lower frequency
b cannot be polarised
c have a higher wavelength
d have a higher frequency.

7 Light of wavelength 640 nm falls on a pair of slits, forming fringes 2.80 mm apart on the screen. What is the new fringe spacing if the wavelength were 400 nm?

a 2.28 mm   b 1.75 mm
c 4.48 mm   d 2.80 mm

8 A ray of light travels from water into air. What happens to the frequency, speed and wavelength of the light?

	Frequency	speed	wavelength
a	increases	increases	decreases
b	decreases	increases	increases
c	unchanged	increases	increases
d	unchanged	increases	decreases

9 What is the ratio of the intensity of two sounds if one is 6.0 dB louder than the other?

a 0.40   b 4.0   c 60   d 10⁶

10 A person has a far point which is 9.5 cm from his eyes. What is the power of the corrective lens needed for him to view an object at infinity?

a −1.1 D   b +1.1 D   c −2.9 D   d +2.9 D

Structured questions

11 a Explain what is meant by simple harmonic motion. [2]
   b Show that the period of oscillation *T* of a mass *m* attached to a spring, having a spring constant *k* is given by $T = 2\pi\sqrt{\dfrac{m}{k}}$. [6]

   c A small particle P undergoes simple harmonic motion. The displacement *x* of P in metres is given by $x = 1.4 \times 10^{-3} \sin(5\pi)\,t.$, where *t* is in seconds.

     i What is the angular frequency of P? [2]
     ii Determine the period of oscillation. [2]
     iii Sketch the velocity–time graph for the motion of P. [3]
     iv Sketch the acceleration time graph for the motion of P. [3]
     v When *t* = 1.0 s determine:
       1 the displacement of P [2]
       2 the velocity of P [2]
       3 the acceleration of P. [2]

12 a State the conditions necessary for the formation of a stationary wave. [2]
   b Explain what is meant by a node and an antinode on a stationary wave. [2]

c   A stationary wave is formed on a stretched string. There are four nodes present. The distance between the four nodes is 1.40 m. The speed of the waves on the string is 220 m s^{-1}.

   i   State the wavelength of the waves on the string. [2]

   ii   Calculate the frequency of vibration. [2]

13 A laser produces light of wavelength 640 nm. When the light is incident normally on a diffraction grating, the first order maximum is produced at 14°.

Calculate:

a   the spacing between the lines on the grating [3]

b   the number of positions of maximum light intensity. [3]

14 a   When a parallel beam of red light is incident normally on a diffraction grating, the light leaving the grating has maxima of intensity in particular directions. Explain the role played by diffraction and interference in producing these maxima. [4]

b   Light consisting of two wavelengths is incident normally on a diffraction grating. The shorter wavelength is 470 nm. At a diffraction angle of 51°, the third order maximum produced by light of this wavelength coincides with the second order maximum produced by light of the other wavelength. Calculate

   i   the spacing between the lines on the grating [2]

   ii   the number of lines per metre on the grating [1]

   iii   the other wavelength present in the light [2]

   iv   the highest order maximum that can be observed with 470 nm wavelength. [2]

15 The spectrum of electromagnetic waves is divided into a number of regions.

a   State three features of waves common to all electromagnetic waves. [3]

b   Arrange the following in increasing magnitude of wavelength:

visible light, X-rays, microwaves, infrared light [4]

c   State a typical wavelength for

   i   red light [1]

   ii   X-rays [1]

   iii   microwaves. [1]

16 The wavelength of monochromatic light can be measured using Young's double slit experiment.

a   Describe an experiment to measure the wavelength of a lamp producing monochromatic red light. State what measurements are taken and explain how these measurements are used to calculate the wavelength of red light. [5]

b   In the experiment you described, state a typical value for:

   i   the width of the slits. [1]

   ii   the distance between the slits [1]

   iii   the distance between the slits and the screen. [1]

c   Explain the role played by diffraction and interference in producing observable fringes. [4]

17 a   State the conditions necessary to obtain observable fringe patterns in the Young's double slit experiment. [3]

b   Derive an expression to show the relationship between the wavelength λ of the light source, the slit separation a, the distance between the slits and the screen D and the spacing between the fringes x. [6]

c   Young's fringes were formed using monochromatic light and two slits with a separation of 0.50 mm. The distance between the screen and the slits is 1.2 m. The distance between ten fringes is 12 mm. Calculate the wavelength of light used. [2]

d   Describe the effect, if any, on the **separation and maximum brightness** of the fringes when the following changes are made in the experiment in **c**.

   i   The distance between the slit and the screen is increased, keeping the slit separation and the light source constant. [2]

   ii   The wavelength of the light source is increased, keeping the slit separation and the distance between the slits and the screen constant. [2]

   iii   The intensity of the light incident on the double slit is increased, keeping the distance between the slits and the screen and the light source constant. [3]

18 a   Explain what is meant by the term *diffraction*. [2]

b   Describe an experiment to demonstrate the diffraction of water waves in a ripple tank. Draw diagrams to illustrate both narrow gap and wide gap diffraction. [6]

c   Explain how diffraction is affected by the width of the gap. [2]

d   Explain why diffraction of sound waves is more easily observed than the diffraction of light. [3]

14 Temperature

14.1 Temperature and temperature scales

Learning outcomes

On completion of this section, you should be able to:

- understand the concept of temperature
- define a thermometric property
- state examples of thermometric properties
- explain how a temperature scale is defined
- explain what is meant by the thermodynamic scale of temperature.

Temperature

Temperature is the measure of the degree of hotness of a body. When two bodies A and B, of different temperatures, are in contact with each other, thermal energy flows between them. If the temperature of body A is higher than the temperature of body B, thermal energy flows from A to B (Figure 14.1.1). Thermal energy will continue to flow until both bodies are at the same temperature. At this stage, both bodies are said to be in **thermal equilibrium** with each other. There is zero net flow of thermal energy between the bodies. However, thermal energy is continuously moving back and forth between the two bodies.

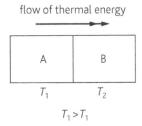

Figure 14.1.1 *Thermal energy flows as a result of a temperature difference*

Defining a temperature scale

In order to define a temperature scale, the concepts of a **thermometric property** and **fixed points** need to be understood. A thermometric property is some property of a material that varies continuously with temperature. The following are examples of thermometric properties.

- The volume of mercury in a capillary tube

  The length of mercury in the capillary tube will vary with temperature. As the temperature of the mercury increases, its volume increases and it moves up the capillary tube. If the temperature of the mercury decreases, its volume decreases. The length of mercury in the capillary tube therefore decreases.

- The electrical resistance of a coil of wire

  The electrical resistance of a metal increases with increasing temperature.

- The pressure of gas contained in a fixed container (volume kept constant)

  If some air is placed inside a container, the pressure inside the container varies with temperature. As the temperature increases, the pressure increases. As the temperature decreases, the pressure decreases.

- The e.m.f. generated when two dissimilar pieces of metals are connected

  When two metals such as copper and iron are connected, an e.m.f. is generated, which is dependent on the temperature of the junction at which the two metals are connected.

A thermometric property should:

- vary smoothly over the range of temperatures being measured
- be sensitive to small changes in temperature
- respond rapidly to changing temperatures.

Once a thermometric property is selected, **fixed points** need to be defined. A fixed point is a standard degree of hotness that can easily be reproduced. For example, the Celsius scale is defined by using two fixed points as follows.

- The **lower fixed point** (ice point) is the temperature of pure melting ice at standard atmospheric pressure. This temperature is defined as $0\,°C$.
- The **upper fixed point** (steam point) is the temperature of steam above pure boiling water at standard atmospheric pressure. This temperature is defined as $100\,°C$.

The chosen thermometric property X, is measured under the two conditions stated above. The two points $(0, X_0)$ and $(100, X_{100})$ are plotted on a graph as shown in Figure 14.1.2. The Celsius scale assumes a linear relationship and therefore a straight line is drawn between the two fixed points. The range is divided up into one hundred equally spaced intervals. In order to determine an unknown temperature θ, the following relationship is used.

Equation

$$\theta = \frac{X_\theta - X_0}{X_{100} - X_0} \times 100\,°C$$

θ – unknown temperature

X_θ – value of thermometric property at temperature θ

X_{100} – value of thermometric property at $100\,°C$

X_0 – value of thermometric property at $0\,°C$

Definition

A thermometric property is a property of a material that varies continuously with temperature.

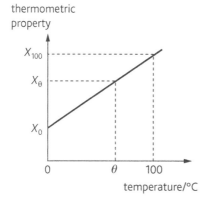

Figure 14.1.2 *Defining a temperature scale*

Thermometers are used to measure temperatures. Each type of thermometer uses a particular thermometric property. If different thermometers are used to measure the temperature of pure melting ice at standard pressure (ice point), all of them would give the same reading. This would also be true at the steam point as well. However, if different thermometers are used to measure the temperature of a glass of water, they may all give different readings. The reason for this is that all the thermometric properties may not vary the same way between the fixed points. The relationship between the thermometric property and temperature may not be linear for all the thermometers.

In order to avoid confusion, a scale was defined that is totally independent of the properties of any particular substance. This scale is called the **thermodynamic scale** or the absolute scale. On this scale, temperatures are measured in kelvin (K). The fixed points used in the thermodynamic scale are:

- Absolute zero (0 K), which is the lowest temperature possible. At this temperature, the internal energy of substances is at a minimum.
- The **triple point of water** (273.16 K). This is the temperature at which ice, water and water vapour are in thermal equilibrium.

The thermodynamic temperature T is given by:

Equation

$$T = \frac{P_T}{P_{tr}} \times 273.16 \, \text{K}$$

T – thermodynamic temperature/K
P_T – pressure of an ideal gas at a temperature T/Pa
P_{tr} – pressure of the same volume of an ideal gas at the triple point of water/Pa

Although the scale is theoretical, it is identical to the scale based on the pressure variation of an ideal gas at constant volume.

The SI unit of temperature is the **kelvin (K)**. An interval of 1 kelvin is defined as being $1/273.16$ of the temperature of the triple point of water as measured on the thermodynamic scale of temperature.

Another commonly used unit, the **degree Celsius (°C)** is defined by the following equation.

Equation

$$\theta = T - 273.15$$

θ – temperature in degrees Celsius/ °C
T – temperature in kelvin/K

It follows that a temperature change of 1 K is exactly equal to a temperature change of 1 °C.

Example

A resistance thermometer is placed in pure water at $0\,°C$ and its resistance is found to be $3825\,\Omega$. At $100\,°C$, its resistance is $185\,\Omega$. The thermometer is placed in a liquid of unknown temperature and the resistance is found to be $976\,\Omega$. A constant-volume gas thermometer is used to measure the unknown temperature and it is found to be $65\,°C$.

i   Calculate the value of the unknown temperature using the resistance thermometer.

ii State your answer in kelvin.

iii Suggest a reason for the difference between the readings on the resistance and constant-volume gas thermometer.

i $\theta = \dfrac{R_\theta - R_0}{R_{100} - R_0} \times 100\,°C$

$ = \dfrac{976 - 3825}{185 - 3825} \times 100$

$ = \dfrac{-2849}{-3640} \times 100$

$\theta = 78.3\,°C$

ii In order to convert from degrees Celsius to kelvin we use:

$\theta = T - 273.15$

$T = \theta + 273.15$

$ = 78.3 + 273.15$

$ = 351.45\,K$

iii The difference in reading is due to the fact that different thermometric properties respond differently and uniquely to the changes in temperature between the fixed points. The assumption is that the thermometric property varies linearly between the two fixed points. This assumption may not be true for different thermometric properties.

Key points

- Temperature is a measure of the hotness of a body.

- A thermometric property is one that varies with temperature.

- A thermometric property can be used to measure temperature. Thermometric properties include – the change in volume of a liquid, the change in resistance of a wire, the change in pressure of a fixed volume of gas and the e.m.f. generated when two dissimilar metals are connected.

- A temperature scale can be defined by choosing a suitable thermometric property and defining fixed points.

- There are two fixed points on the Celsius scale (ice point and steam point).

- There are two fixed points on the thermodynamic scale (absolute zero and the triple point of water).

- The thermodynamic scale is called an absolute scale because it does not depend on any thermometric property and is theoretical.

- A change of one degree Celsius is exactly equal to a change of one kelvin.

Learning outcomes

On completion of this section, you should be able to:

- describe the principal features of a liquid-in-glass, a resistance, a thermistor, a thermocouple and a constant-volume gas thermometer

- state the advantages and disadvantages of various types of thermometers

- select an appropriate thermometer to measure temperature.

Types of thermometers

Liquid-in-glass thermometer

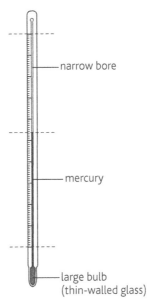

Figure 14.2.1 *A liquid-in-glass thermometer*

In a liquid glass-in thermometer, a thin walled glass bulb is filled with mercury. When heated, the change in volume of the mercury causes it to rise up the fine bore of the thermometer (Figure 14.2.1). These thermometers are inexpensive and easy to use. Glass is used because it is transparent and does not expand more than the mercury. Mercury is used because it is opaque. This means that it is easily seen through the glass. Mercury does not wet glass. This means that it does not stick to the glass while moving up or down the bore. Another advantage of using mercury is that it is a metal and is a very good conductor of heat.

Mercury freezes at 234 K, which means that it cannot be used to measure very low temperatures. Mercury-in-glass thermometers have relatively large thermal heat capacities. This means that they cannot respond quickly to rapidly changing temperatures. It also means that the thermometer will affect the temperature being measured. The reading will therefore be lower than the actual value. In order to make this thermometer sensitive, the bore is made very thin. The drawback to this is that it limits the range of the thermometer of a given length.

Resistance thermometers

In a platinum-resistance thermometer, a platinum wire is coiled on an insulating material like mica (Figure 14.2.2). It relies on the fact that the resistance of the platinum wire varies with temperature. Platinum is used because it has a high temperature coefficient of resistance. This means that there is a large change in resistance for a small change in temperature. This makes the thermometer very sensitive. Platinum has a melting point of 2046 K. This means that the thermometer has a very large range. A Wheatstone bridge is used to measure the change in resistance of the platinum. Since a Wheatstone bridge is used, very slight changes in

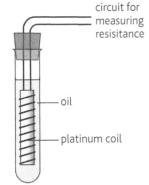

Figure 14.2.2 *A platinum-resistance thermometer*

resistance can be measured very accurately. The biggest drawback of this type of thermometer is that it has a relatively large heat capacity. This means that it is slow to respond to rapidly changing temperatures.

Thermistors

A **thermistor** is a non-linear device with a thermally sensitive resistor. The device is very small. The resistance of the thermistor varies with temperature. A thermistor has a negative temperature coefficient, which means that its resistance decreases with increasing temperature. Thermistors have very small heat capacities. This means that they can respond quickly to rapidly changing temperatures. It also means that they have little effect on the temperatures being measured. The disadvantage of this type of thermometer is that it is less stable than a platinum-resistance thermometer and needs periodic calibration. This makes it less accurate when used to measure temperatures over a long period of time.

Thermocouples

A **thermocouple** makes use of the **thermoelectric effect** or the Seebeck effect. When two different metals are joined together as in Figure 14.2.3, an e.m.f. is produced and an electric current flows in the circuit. The value of the e.m.f. generated is dependent on the metals being used and the temperature difference between the junctions J_1 and J_2. The variation of the e.m.f. with temperature of J_2 is almost always parabolic, when J_1 is kept at $0\,^\circ C$. In order to measure an unknown temperature, one junction J_1 is placed in melting ice, which provides the reference temperature, while the other junction J_2 is placed in contact with the object of unknown temperature. Thermocouples have very small junctions. They therefore have very small heat capacities. This means that they can respond quickly to rapidly changing temperatures.

Figure 14.2.4 shows the variation of e.m.f. with temperature for a thermocouple. The thermocouple is less sensitive in the region just below and above the neutral temperature (temperature that produces the maximum e.m.f.). The change in e.m.f. as a result of a temperature change is very small in this region. Also, it can be seen from the graph, that for a given e.m.f. there are two possible values for the temperature being measured.

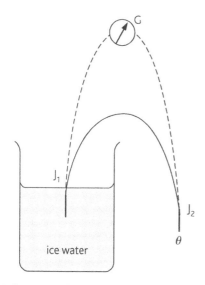

Figure 14.2.3 A thermocouple

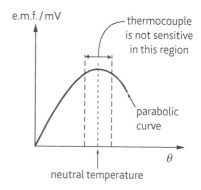

Figure 14.2.4 Variation of e.m.f. with temperature for a thermocouple

Constant-volume gas thermometers

In this type of thermometer, the pressure of a fixed mass of gas is measured using a mercury manometer (Figure 14.2.5). The glass bulb holds a fixed volume of gas. In order to ensure that the volume of the gas remains constant, the height of the movable tube is adjusted so that the mercury is always at the fixed volume marker. When measuring temperature, the glass bulb is brought in contact with the substance being measured. The gas expands and pushes the gas up the movable tube. The height of the movable tube is adjusted so that the mercury on the left hand of the tube returns to the fixed volume mark. The pressure of the gas is then measured using

$$p_\theta = p + \rho h g$$

where p_θ is the pressure of the gas at a temperature θ, p is atmospheric pressure, g is the Earth's **gravitational field strength** ($9.81\,\mathrm{N\,kg^{-1}}$), ρ is the density of mercury and h is the height of the mercury in the U-tube.

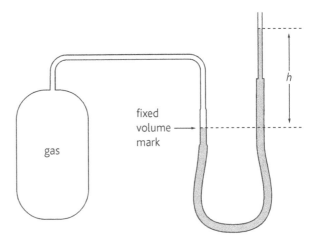

Figure 14.2.5 *A constant-volume gas thermometer*

Table 14.2.1 compares the various thermometers that have been discussed.

Table 14.2.1 *Comparison of different types of thermometers*

Type of thermometer	Range/K	Thermometric property	Advantages	Disadvantages
Mercury-in-glass thermometer	234–710	Volume of mercury in a fine column	▪ Easy to use ▪ Portable	▪ Limited range ▪ Fragile
Resistance thermometer	25–1750	Electrical resistance of a platinum coil	▪ Accurate ▪ Wide range ▪ Can measure small temperature differences	▪ Slow response (not suitable for rapidly changing temperatures) ▪ Not a direct reading
Thermocouple	80–1400	e.m.f. across the junction of two dissimilar metals	▪ Wide range ▪ Small ▪ Fast response ▪ Can take remote readings and send to a computer	▪ Not as accurate as a resistance or constant-volume thermometer
Constant-volume gas thermometer	3–1750	Pressure of a fixed mass of gas at constant volume	▪ Wide range ▪ Accurate	▪ Bulky ▪ Slow response ▪ Fragile

In order to measure the temperature of a substance, the choice of thermometer would depend on:

- the ease of use of the thermometer
- the range of the thermometer
- the response time of the thermometer
- the accuracy of the thermometer.

Example

Select a suitable choice of thermometer to measure the following temperatures.

a The temperature of a very small quantity of ethanol.

b The boiling point of oxygen (90 K).

c A small change in body temperature.

d A rapidly changing temperature.

a For a very small quantity of ethanol, a thermometer with a very small heat capacity will have to be used. The reason for this is that the thermometer will lower the temperature of the ethanol and will not give its true temperature. The most suitable thermometer would be the thermocouple, since it has small junctions and therefore has a very small heat capacity.

b A platinum-resistance thermometer would be suitable in this case. Boiling of the oxygen occurs at a constant temperature. Platinum-resistance thermometers are very accurate and can measure steady temperatures.

c A mercury-in-glass thermometer would be suitable in this case. Body temperature is around 37 °C.

d The best thermometer would be the thermocouple. Thermocouples have small heat capacities and therefore respond quickly to rapidly changing temperatures.

Key points

- There are advantages and disadvantages to using the following thermometers:
  - liquid-in-glass
  - resistance
  - thermistor
  - thermocouple
  - constant-volume gas thermometer.

Learning outcomes

On completion of this section, you should be able to:

- define the term *internal energy*
- relate a rise in temperature of a body to an increase in internal energy
- define the terms *heat capacity* and *specific heat capacity*.

Definition

The internal energy of a system is the sum of the random distribution of the kinetic and potential energies of the particles that make up the system.

Internal energy

Substances are made up of many particles. Consider a piece of iron being heated in a Bunsen flame. The particles inside the iron gain thermal energy and vibrate about their mean positions. Their kinetic energy increases. If more energy is supplied, the particles move further apart from each other. Their potential energy increases. So as thermal energy is supplied to the iron, the kinetic and potential energy of the particles increases. The **internal energy** of the piece of iron is the sum of the random distribution of the kinetic and potential energy of the particles.

Internal energy $\qquad U = E_K + E_P$

In a gas there are many molecules present. The molecules are constantly in motion. This means that the molecules possess kinetic energy. The motion also causes the molecules to move further apart (against the attractive forces that exist between the molecules). Therefore, the molecules possess potential energy as well. The internal energy of the gas is the sum of the random kinetic and potential energies of the molecules that are present.

The internal energy of a system is determined by the state of the system.

Whenever energy is supplied to a system, the particles within it begin vibrating more rapidly. The kinetic energy of the particles increases. The potential energy of the particles also increases. Consequently, the internal energy of the system increases.

In an **ideal gas**, the force of attraction between the molecules is negligible. Therefore, the internal energy is only dependent on the kinetic energies of the molecules that make up the gas.

There are two ways to increase the internal energy of a gas:

- supply energy to the gas (heating it)
- compress the gas (do work on it).

Suppose there is a quantity of gas inside a metal container. Energy can be supplied to the gas by heating the container. The particles that make up the container begin to vibrate rapidly. When the gas molecules strike the walls of the container they bounce off with a greater speed. The average kinetic energy of all the particles that make up the gas increases. The result is that the internal energy and the temperature of the gas increases.

If the gas inside a cylinder is compressed using a piston, the temperature of the gas increases. When the gas molecules strike the moving piston, they bounce off with a greater speed. The average kinetic energy of all the particles that make up the gas increases. The result is that the internal energy and the temperature of the gas increase.

Heat capacity and specific heat capacity

The amount of energy required to increase the temperature of a substance depends on the mass of the substance, the required temperature rise and the material itself. Some materials are easier to heat up than others.

The amount of energy needed to raise the temperature of 1 kg of a substance by 1 °C (or 1 K) is called the **specific heat capacity** (c) of the substance.

The amount of energy E_H required to produce a temperature change of $\Delta\theta$ in a substance of mass m kg is given by

$$E_H = mc\Delta\theta$$

where c is the specific heat capacity of the substance.

The SI unit of specific heat capacity is $c = \dfrac{E_H}{m\Delta\theta} = \dfrac{J}{kgK} = J\,kg^{-1}\,K^{-1}$

The unit can also be expressed as $J\,kg^{-1}\,°C^{-1}$. This is so because a change in 1 °C is exactly equal to a change in 1 K.

The larger the specific heat capacity of the substance, the more energy that is required to increase its temperature. The specific heat capacity of aluminium and copper are $880\,J\,kg^{-1}\,K^{-1}$ and $380\,J\,kg^{-1}\,K^{-1}$ respectively. Suppose 2 kg of aluminium and copper are heated to produce a temperature change of 10 °C.

The amount of energy absorbed by the aluminium is given by

$E_H = mc\Delta\theta = 2 \times 880 \times 10 = 17\,600\,J$

The amount of energy absorbed by the copper is given by

$E_H = mc\Delta\theta = 2 \times 380 \times 10 = 7\,600\,J$

This simple calculation shows that the higher the specific heat capacity, the more energy that is required to raise the temperature of one kilogram of the substance.

The same equation can be used to determine the amount of energy released when a substance cools. Suppose 2 kg of aluminium cools by 10 °C. The amount of energy released is 17 600 J.

Another useful quantity is **heat capacity (C)**. The heat capacity of a substance is the energy required to raise the temperature of a substance by one degree.

The unit of heat capacity is the $J\,K^{-1}$ or $J\,°C^{-1}$.

The amount of energy E_H required to produce a temperature change of $\Delta\theta$ in a substance is given by $E_H = C\Delta\theta$.

Key points

- The internal energy of a system is the sum of the random distribution of the kinetic and potential energies of the particles that make up the system.
- The internal energy of an ideal gas is dependent on the kinetic energy of the molecules only.
- The internal energy of a gas can be increased by supplying energy to the gas or doing work on the gas (compressing the gas).
- The specific heat capacity of a substance is the amount of energy required to raise the temperature of 1 kg of a substance by 1 °C or 1 K.
- The SI unit of specific heat capacity is the $J\,kg^{-1}\,K^{-1}$.
- The heat capacity of a substance is the amount of energy required to raise the temperature of a substance by 1 °C.

Definition

The specific heat capacity of a substance, c, is the amount of heat energy required to increase the temperature of one kilogram of a substance by one degree.

Equation

$E_H = mc\Delta\theta$

E_H – amount of energy absorbed
m – mass/kg
c – specific heat capacity/$J\,kg^{-1}\,K^{-1}$ or $J\,kg^{-1}\,°C^{-1}$
$\Delta\theta$ – change in temperature/K or °C

Definition

The heat capacity of a substance, C, is the amount of heat energy required to increase the temperature of a substance by one degree.

Equation

$E_H = C\Delta\theta$

E_H – amount of energy absorbed
C – heat capacity/$J\,K^{-1}$ or $J\,°C^{-1}$
$\Delta\theta$ – change in temperature/K or °C

Equation

$C = mc$

C – heat capacity/$J\,K^{-1}$
m – mass/kg
c – specific heat capacity/$J\,kg^{-1}\,K^{-1}$

Learning outcomes

On completion of this section, you should be able to:

- describe experiments to determine specific heat capacities of substances by electrical methods.

Electrical methods for measuring specific heat capacities

Specific heat capacities of substances can be experimentally determined by using an electrical method or by a method of mixtures. This section deals with the electrical methods of determining the specific heat capacity of liquids and solids.

Specific heat capacity of a liquid

The apparatus used to measure the specific heat capacity of liquid is shown in Figure 15.2.1. A known mass of liquid m is placed in a calorimeter (the vessel in which the heat measurements are made). The initial temperature of the liquid θ_1 is measured using the thermometer. An electric current is then made to flow through the heating element for a length of time t. The ammeter and voltmeter reading I and V are recorded. During the heating period the stirrer is used to provide uniform heating of the liquid. The final temperature θ_2 is recorded.

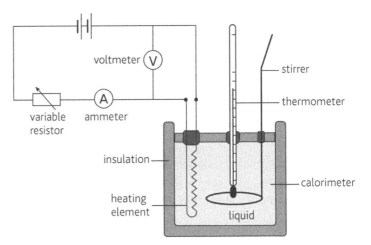

Figure 15.2.1 *An electrical method to determine the specific heat capacity of a liquid*

Experimental results

Mass of liquid $= m\,kg$

Initial temperature of liquid $= \theta_1\,°C$

Final temperature of liquid $= \theta_2\,°C$

Ammeter reading $= I\,amps$

Voltmeter reading $= V\,volts$

Time heating element was switched on $= t\,seconds$

Electrical energy supplied to liquid $= IVt$

Energy absorbed by the liquid $= mc(\theta_1 - \theta_2)$

where c is the specific heat capacity of the liquid.

If it is assumed that no heat losses occur, the specific heat capacity of the liquid can be found by using the following:

Electrical energy supplied by heating element = energy absorbed by liquid

$$IVt = mc(\theta_1 - \theta_2)$$

$$\therefore \quad c = \frac{IVt}{m(\theta_1 - \theta_2)}$$

Specific heat capacity of a solid

The apparatus used to measure the specific heat capacity of a solid is similar to the one used in the previous experiment. Two holes are drilled in the solid. The mass of solid m is determined using an electronic balance. A thermometer is placed in one hole and a heating element is placed in the second hole. A small amount of oil is placed inside the hole with the thermometer to ensure that there is good thermal contact between the thermometer and the solid. The initial temperature of the solid θ_1 is measured using the thermometer. An electric current is then made to flow through the heating element for a length of time t. The ammeter and voltmeter reading I and V are recorded. The final temperature θ_2 is recorded (Figure 15.2.2).

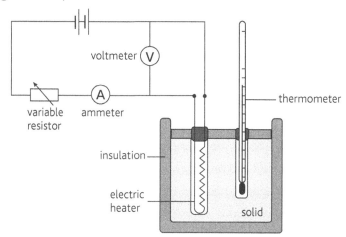

Figure 15.2.2 *An electrical method to determine the specific heat capacity of a solid*

Experimental results

Mass of solid = m kg

Initial temperature of solid = θ_1 °C

Final temperature of solid = θ_2 °C

Ammeter reading = I amps

Voltmeter reading = V volts

Time heating element was switched on = t seconds

Electrical energy supplied to solid = IVt

Energy absorbed by the solid = $mc(\theta_1 - \theta_2)$

where c is the specific heat capacity of the solid.

If it is assumed that no heat losses occur, the specific heat capacity of the solid can be found by using the following:

Electrical energy supplied by heating element = energy absorbed by solid

$$IVt = mc(\theta_1 - \theta_2)$$

$$\therefore \quad c = \frac{IVt}{m(\theta_1 - \theta_2)}$$

A continuous flow method to determine the specific heat capacity of a liquid

In 1899 Callendar and Barnes devised a method to measure the specific heat capacity of a liquid by a continuous flow method.

The liquid is allowed to flow from a constant-head tank through a glass tube containing a heating element as shown in Figure 15.2.3.

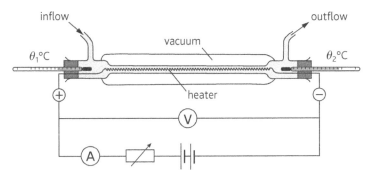

Figure 15.2.3 *Continuous flow method*

The glass tube is insulated by a glass jacket, which is evacuated, so that heat cannot escape from the liquid. The experiment is allowed to reach a steady state. At this stage, the temperature of the liquid entering and leaving the glass tube remains constant. A suitable thermometer for this experiment is a platinum-resistance thermometer. It is extremely accurate and is good at measuring steady temperatures.

Temperature of liquid entering glass tube $= \theta_1 \,°C$

Temperature of liquid leaving glass tube $= \theta_2 \,°C$

Mass of water collected as it leaves the tube $= m_1 \, kg$

Time taken to collect liquid $= t$ seconds

Ammeter reading $= I_1$ amps

Voltmeter reading $= V_1$ volts

Under steady state conditions, all the energy supplied by the heating element is carried away by the flowing liquid.

$$I_1 V_1 t = m_1 c (\theta_1 - \theta_2) + h \qquad (1)$$

where h is the heat lost to the surroundings in a time t and c is the specific heat capacity of the liquid.

The rate at which the liquid is flowing into the glass tube is altered so that the amount collected in a time t is m_2. The current and voltage are adjusted to bring the temperature θ_2 back to its original value. The temperature of the liquid entering the glass tube is the temperature of the liquid leaving the constant-head tank. This temperature θ_1 is constant. Since all the temperatures are the same, the heat lost in a time t is again h.

Temperature of liquid entering glass tube $= \theta_1 \,°C$

Temperature of liquid leaving glass tube $= \theta_2 \,°C$

Mass of water collected as it leaves the tube $= m_2 \, kg$

Time taken to collected liquid $= t$ seconds

Ammeter reading $= I_2$ amps

Voltmeter reading $= V_2$ volts

Under steady state conditions, all the energy supplied by the heating element is carried away by the flowing liquid.

$\therefore \qquad I_2 V_2 t = m_2 c (\theta_2 - \theta_1) + h \qquad\qquad\qquad (2)$

where h is the heat lost to the surroundings in a time t.

Subtracting Equation (1) from Equation (2):

$$(I_2 V_2 - I_1 V_1)t = c(m_2 - m_1)(\theta_2 - \theta_1)$$

$$c = \frac{(I_2 V_2 - I_1 V_1)t}{(m_2 - m_1)(\theta_2 - \theta_1)}$$

The advantages of the continuous flow method are:

- The inlet and outlet temperatures can be measured very accurately using a platinum-resistance thermometer.
- The heat lost in the experiment is not required since it is eliminated in the calculations. Therefore, the heat capacity of various parts of the apparatus is not required.

The main disadvantage of this method is that a large volume of the liquid is required.

Key points

- The specific heat capacity of a liquid and a solid can be determined by an electrical method.

15.3 Measuring specific heat capacities by using a method of mixing

Learning outcomes

On completion of this section, you should be able to:

- describe experiments to measure specific heat capacities of solids and liquids by using a method of mixtures.

Measuring the specific heat capacity of a solid

The specific heat capacity of a solid can be determined by mixing it with a liquid of known specific heat capacity.

Suppose you are required to determine the specific heat capacity c of a small piece of metal. The mass of the metal is measured using an electronic balance. The metal is attached to a piece of thread. The metal is then placed into a container of boiling water. It is left there for approximately 10 minutes to ensure that the metal and the water are at the same temperature. The temperature of the boiling water is measured using a thermometer.

In the meantime, some cold water is placed inside a polystyrene cup. The mass of the water is measured. The initial temperature of the water is measured. The hot piece of metal is quickly added to the cold water. A stirrer is used to stir the water gently. The highest temperature achieved by the water is recorded.

Mass of piece of metal $= m_1$ kg

Initial temperature of the metal $= \theta_1\,°C$

Mass of cold water $= m_2$ kg

Initial temperature of cold water $= \theta_2\,°C$

Final temperature of cold water $= \theta_3\,°C$

Specific heat capacity of water $= c_w\,J\,kg^{-1}\,K^{-1}$

Heat lost by piece of metal $=$ heat gained by water

$$m_1 \times c \times (\theta_1 - \theta_3) = m_2 \times c_w \times (\theta_3 - \theta_2)$$

$$c = \frac{m_2 \times c_w \times (\theta_3 - \theta_2)}{m_1 \times (\theta_1 - \theta_3)}$$

Measuring the specific heat capacity of a liquid

The specific heat capacity of a liquid can be determined by mixing it with a liquid or solid of known specific heat capacity. The calculation is similar to that used above.

Example

Calculate the energy required to:

a   Increase the temperature of 0.75 kg of aluminium from 25 °C to 60 °C.

b   Increase the temperature of 1.2 kg of copper from 10 °C to 45 °C.

[Specific heat capacity of aluminium and copper are $880\,J\,kg^{-1}\,K^{-1}$ and $380\,J\,kg^{-1}\,K^{-1}$ respectively]

a $E_H = mc\Delta\theta = 0.75 \times 880 \times (60 - 25) = 23\,100\,J$

b $E_H = mc\Delta\theta = 1.2 \times 380 \times (45 - 10) = 15\,960\,J$

Example

Calculate the energy released when 1.5 kg of water cools from 90 °C to 25 °C.

[Specific heat capacity of water = 4200 J kg^{-1} K^{-1}]

$E_H = mc\Delta\theta = 1.5 \times 4200 \times (90 - 25) = 409\,500$ J

Example

A kettle rated at 2 kW is used to raise the temperature of 1.5 kg of water from 30 °C to 100 °C.

Calculate:

a the energy required to raise the temperature of the water from 30 °C to 100 °C

b the time taken to raise the temperature of the water from 30 °C to 100 °C.

[Specific heat capacity of water = 4200 J kg^{-1} K^{-1}]

a Energy required $E_H = mc\Delta\theta = 1.5 \times 4200 \times (100 - 30) = 4.41 \times 10^5$ J

b Energy = power × time

$t = \dfrac{E}{P} = \dfrac{4.41 \times 10^5}{2 \times 10^3} = 220.5$ s

Key points

- The specific heat capacity of a liquid or solid can be determined by mixing with another substance (solid or liquid) of known specific heat capacity.

Change of state

Matter exists in one of three states – solids, liquids and gases. When energy is supplied to a solid, it changes into a liquid and eventually into a gas. Inside a solid, there are many particles held together by strong bonds. When a solid is heated, these particles gain energy. The particles begin to vibrate more rapidly about their mean positions. The kinetic energy of the particles increases, resulting in an increase in temperature of the solid. The particles also begin moving further apart and their potential energy increases. Eventually, enough energy is absorbed by the solid, causing the bonds between the particles to break. The motion of the particles becomes more disordered. The particles move freely within the structure. At this point in the heating process, the solid has melted and has become a liquid.

As more energy is supplied to the liquid, the particles vibrate even more rapidly. The kinetic energy of the particles increases, resulting in an increase in temperature of the liquid. The particles also begin moving further apart and their potential energies increase. Eventually enough energy is absorbed by the liquid to cause the bonds between the particles to break. At this stage, the particles are much further apart, moving rapidly and randomly and in a disordered state. At this point in the heating process, the liquid has become a gas (Figures 15.4.1 and 15.4.2).

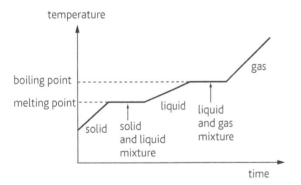

Figure 15.4.1   *A heating curve (solid to liquid to gas)*

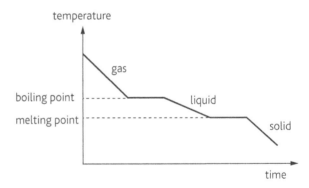

Figure 15.4.2   *A cooling curve*

At a change of state, all the energy supplied is used to break the bonds between the particles that make up the substance. All the energy supplied increases the potential energy of the particles. The energy supplied does not increase the kinetic energy of the particles during a change of state. **Since the temperature of a substance is dependent on the kinetic energy of the particles, it follows that at a change of state, there is no change in temperature.**

Melting is the process whereby a solid changes into a liquid without a change in temperature.

Boiling is the process whereby a liquid changes into a gas (vapour) without a change in temperature.

Latent heat

The energy required to change the state of a substance is known as the **latent heat**.

When a solid changes into a liquid, latent heat is required. The **specific latent heat of fusion** is the energy required to convert unit mass (1 kg) of a substance from a solid into liquid without a change in temperature.

The SI unit of specific latent heat of fusion is the Jkg^{-1}. The specific latent heat of fusion is calculated using the following equation.

Equation

$E_H = ml_f$

E_H – energy/J
m – mass/kg
l_f – specific latent heat of fusion/Jkg^{-1}

Example

The specific latent heat of fusion of ice is $3.3 \times 10^5 Jkg^{-1}$. Calculate the amount of energy required to convert 80 g of ice at 0°C into water at 0°C.

$$E_H = ml_f = 0.08 \times 3.3 \times 10^5 = 2.64 \times 10^4 J$$

The **specific latent heat of vaporisation** is the energy required to convert unit mass (1 kg) of a substance from a liquid into a vapour without a change in temperature.

The SI unit of specific latent heat of vaporisation is the Jkg^{-1}. The specific latent heat of vaporisation is calculated using the following equation.

Equation

$E_H = ml_v$

E_H – energy/J
m – mass/kg
l_v – specific latent heat of vaporisation/Jkg^{-1}

Example

The specific latent heat of vaporisation of water is $2.3 \times 10^6 Jkg^{-1}$. Calculate the amount of energy required to convert 250 g of water at 100°C to steam at 100°C.

$$E_H = ml_v = 0.25 \times 2.3 \times 10^6 = 5.75 \times 10^5 J$$

Example

Calculate the amount of energy required to convert 1.2 kg of ice at −10°C to steam at 100°C.

Specific heat capacity of ice $= 2.1 \times 10^3 Jkg^{-1}K^{-1}$

Specific latent heat of fusion of ice $= 3.3 \times 10^5 Jkg^{-1}$

Specific heat capacity of water $= 4.2 \times 10^3 Jkg^{-1}$

Specific latent heat of vaporisation of water $= 2.3 \times 10^6 Jkg^{-1}$

Definition

The **specific latent heat of fusion** l_f is the energy required to convert 1 kg of substance from a solid to a liquid without a change in temperature.

Definition

The **specific latent heat of vaporisation** l_v, is the energy required to convert 1 kg of substance from a liquid to a vapour without a change in temperature.

If it is required to determine the energy needed to change the temperature of a substance without a change of state, use the equation $E_H = mc\Delta\theta$.

If it is required to determine the energy needed to change the state of a substance, use the equation $E_H = ml$.

Energy required to raise the temperature of ice to $0\,°C$

$$E_H = mc\Delta\theta = 1.2 \times 2.1 \times 10^3 \times (0 - -10) = 25\,200\,J$$

Energy required to convert ice at $0\,°C$ to water at $0\,°C$

$$E_H = ml_f = 1.2 \times 3.3 \times 10^5 = 396\,000\,J$$

Energy required to raise the temperature of water from $0\,°C$ to $100\,°C$

$$E_H = mc\Delta\theta = 1.2 \times 4.2 \times 10^3 \times (100 - 0) = 504\,000\,J$$

Energy required to convert water at $100\,°C$ to steam at $100\,°C$

$$E_H = ml_v = 1.2 \times 2.3 \times 10^6 = 2\,760\,000\,J$$

Energy required to convert $1.2\,kg$ of ice at $-10\,°C$ to steam at $100\,°C$

$$= 25\,200 + 396\,000 + 504\,000 + 2\,760\,000 = 3.69 \times 10^6\,J$$

The specific latent heat of fusion of ice is $3.3 \times 10^5\,J\,kg^{-1}$ and the specific latent heat of vaporisation of water is $2.3 \times 10^6\,J\,kg^{-1}$. It can be seen that the specific latent heat of vaporisation is much larger than the specific latent heat of fusion of ice. The difference between these two quantities can be attributed to the differences in structures of solids, liquids and gases. In ice (solid water) the particles are held tightly together by strong intermolecular forces. The particles are free to vibrate about their mean positions. In liquid water, the molecules are held together less tightly by intermolecular forces and the separation between them is similar to that of a solid. The molecules are free to move within the body of the liquid. In steam (vaporised water), the molecules are so far apart that the force of attraction between them are negligible. The energy required to completely separate the molecules when changing from a liquid to a gas is much greater than when changing from a solid to liquid, where the mean separation is only increased slightly. Hence the reason why the latent heat of vaporisation is greater that the latent heat of fusion.

Evaporation and boiling

When energy is supplied to a liquid, its temperature rises. If more energy is supplied, the liquid will eventually start to boil. At this point, the temperature of the liquid remains constant (i.e. its boiling point). Bubbles are seen in the boiling liquid. The energy supplied causes the liquid to be converted into a vapour. Table 15.4.1 compares evaporation and boiling.

Boiling – A substance absorbs energy and changes state from a liquid into a gas without a change in temperature.

Suppose the temperature of a room is $25\,°C$. If some water is spilled on the floor in the room, it eventually evaporates. The boiling point of pure water is $100\,°C$, yet the water changes state from liquid into a gas without reaching its boiling point.

Evaporation is the process by which a liquid changes into a gas without reaching its boiling point. Evaporation occurs at any temperature.

Factors that affect the rate of evaporation are:

- temperature (rate increases with increasing temperature)
- surface area of liquid exposed to the atmosphere (rate increases if a larger area is exposed to the atmosphere)
- flow of air above the surface of the liquid (rate increases when air is blown above the surface).

Evaporation plays an important role in cooling the human body. Suppose an athlete runs a 100 m race. During the race, the body temperature increases. In order for the body to cool down, sweat is produced. The sweat covers the body. Thermal energy is required to convert the sweat into vapour. This thermal energy (latent heat) is taken from the surface of the skin as the sweat evaporates. The evaporation of the sweat causes the body to cool.

Table 15.4.1 *Comparing evaporation and boiling*

Evaporation	Boiling
Change of state occurs	Change of state occurs
Occurs at the surface of the liquid	Occurs within the body of the liquid
No bubbles are seen	Bubbles are seen rising
Occurs at any temperature	Occurs at a specific temperature (boiling point)

Using the kinetic model to explain why the remaining liquid in a container cools as it evaporates

The molecules in a liquid have a range of speeds and therefore a range of kinetic energies. Some molecules have sufficient energy to break free from the surface of the liquid. Since the most energetic molecules escape from the surface, the average kinetic energy of the remaining molecules decreases. The temperature of substance is dependent on the average kinetic energy of the particles that make up the substance. Therefore, it can be concluded that the temperature of the remaining liquid decreases as some of it evaporates (Figure 15.4.3).

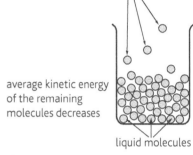

molecules having sufficient energy escape from the surface of the liquid

average kinetic energy of the remaining molecules decreases

liquid molecules

Figure 15.4.3 *Cooling effect associated with evaporation*

Key points

- Melting is the process by which a solid changes into a liquid without a change in temperature.
- Boiling is the process by which a liquid changes into a gas without a change in temperature.
- Latent heat is the energy required to convert a substance from a solid into liquid or liquid into gas without a change in temperature.
- At a change of state, there is no change in temperature of the substance. All the energy supplied is used to break the bonds in the substance.
- The specific latent heat of fusion of a substance is the energy required to change unit mass of a solid into a liquid without a change in temperature.
- The specific latent heat of vaporisation of a substance is the energy required to convert unit mass of a liquid into a vapour without a change in temperature.
- Evaporation is the process whereby a liquid changes into a vapour without reaching its boiling point.

Learning outcomes

On completion of this section, you should be able to:

- describe an experiment to measure the specific latent heat of fusion of a solid

- describe an experiment to measure the specific latent heat of vaporisation of a liquid.

Measuring the specific latent heat of fusion of a solid (ice)

The specific latent heat of fusion of ice can be found by a method of mixtures. The mass of a calorimeter is measured using an electronic balance. Some water which is about 15°C above room temperature is placed inside a calorimeter. The mass of the water and the calorimeter is measured. The specific heat capacities of the water and the calorimeter are known. The initial temperature of the water is measured using a thermometer. Some pieces of ice are blotted until they are dry and then placed gently into the water. A stirrer is used to stir the mixture until all the ice melts. The lowest temperature of the water is then recorded. The mass of the water and the calorimeter is measured again. This will be used to determine the mass of the ice that was added.

The specific latent heat of fusion of ice is determined as follows:

Temperature of water $= \theta_1$ °C

Mass of calorimeter $= m_1$ kg

Mass of calorimeter and water $= m_2$ kg

Mass of water $= m_3 = (m_2 - m_1)$ kg

Mass of water and calorimeter at end of experiment $= m_4$ kg

Mass of ice used $= m_5 = (m_4 - m_2)$ kg

Final temperature of water $= \theta_2$ °C

Specific heat capacity of water $= 4200 \, \mathrm{J\,kg^{-1}\,K^{-1}}$

Specific heat capacity of calorimeter $= c \, \mathrm{J\,kg^{-1}\,K^{-1}}$

$$\begin{array}{l}\text{Energy lost} \\ \text{by water}\end{array} + \begin{array}{l}\text{Energy lost by} \\ \text{calorimeter}\end{array} = \begin{array}{l}\text{Energy used} \\ \text{to melt ice}\end{array} + \begin{array}{l}\text{Energy used to increase} \\ \text{temperature of melted ice}\end{array}$$

$$\therefore \quad m_3 \times 4200 \times (\theta_1 - \theta_2) + m_1 \times c \times (\theta_1 - \theta_2) = m_5 l_f + m_5 \times 4200 \times \theta_2$$

The specific latent heat of fusion can be calculated from the expression above.

Measuring the specific latent heat of vaporisation of a liquid

The apparatus used to measure the specific latent heat of vaporisation of a liquid is shown in Figure 15.5.1. The liquid is heated at a constant rate until it begins boiling. The vapour produced passes through the holes in the inner walls of the jacket and passes through the condenser. The liquid is allowed to boil until all parts of the apparatus become steady. The receiving vessel is then used to collect the liquid. The mass of the liquid collected over a period of time t is determined.

Mass of liquid collected $= m_1$ kg

Current $= I_1$ amps

Voltage $= V_1$ volts

Time $= t$ seconds

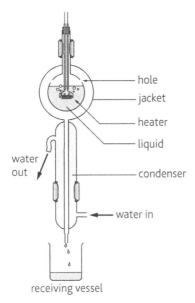

Figure 15.5.1 *Measuring specific latent heat of vaporisation*

Electrical energy supplied = energy required to vaporise liquid + heat losses

$$I_1V_1t = m_1l_v + h \qquad (1)$$

The experiment is repeated, but this time the potential difference across the heater is changed and the new mass m_2 of vapour which condenses in the same time t is measured. The heat loss h in time t is the same as before. This is because heat loss is dependent on the temperature of the liquid, which remains fixed at its boiling point.

Mass of liquid collected = m_2 kg

Current = I_2 amps

Voltage = V_2 volts

Time = t seconds

$$I_2V_2t = m_2l_v + h \qquad (2)$$

Equation (1) – Equation (2) gives

$$I_1V_1t - I_2V_2t = m_1l_v - m_2l_v$$

$$t(I_1V_1 - I_2V_2) = (m_1 - m_2)l_v$$

$$l_v = \frac{(I_1V_1 - I_2V_2)t}{(m_1 - m_2)}$$

Revision questions 7

Answers to questions that require calculation can be found on the accompanying CD.

1   a   Outline how a physical property which varies with temperature may be used to measure temperature. [3]
    b   Suggest why the thermodynamic scale of temperature is called an absolute scale. [2]

2   a   Explain what is meant by a thermometric property. [1]
    b   State three thermometric properties. [3]
    c   Suggest why, on the Celsius scale, different types of thermometers agree at the fixed points but may not necessarily agree at temperatures between the fixed points. [3]

3   A resistance thermometer is placed in pure water at $0\,°C$ and its resistance is found to be $3625\,\Omega$. At $100\,°C$, its resistance is $160\,\Omega$. The thermometer is placed in a liquid of unknown temperature and the resistance is found to be $850\,\Omega$.

    A constant-volume gas thermometer is used to measure the unknown temperature and is found to be $65\,°C$.
    a   Calculate the value of the unknown temperature using the resistance thermometer. [3]
    b   State your answer in kelvin. [2]
    c   Suggest a reason for the difference between the readings on the resistance and constant-volume gas thermometer. [2]

4   Describe the principal features of a thermocouple. [4]

5   State two advantages of a thermocouple over a platinum-resistance thermometer. [2]

6   Briefly describe the features of a constant-volume gas thermometer. [4]

7   State a suitable choice of thermometer to measure the following:
    a   the temperature of a blast furnace [1]
    b   the temperature of various positions in the flame of a Bunsen burner [1]
    c   the temperature of boiling water [1]
    d   the temperature of a room. [1]

8   Discuss whether a thermistor is a suitable choice for measuring the temperature of a laboratory over a period of one year. [3]

9   a   Describe the features of a mercury-in-glass thermometer. [3]
    b   Give one reason why glass is used in this type of thermometer. [1]
    c   State two reasons why mercury is used. [2]
    d   State one advantage and one disadvantage of a mercury-in-glass thermometer over a platinum-resistance thermometer over the same range. [2]

10 a   Explain what is meant by the internal energy of a substance. [2]
    b   A piece of copper is heated in a Bunsen flame. Explain what happens to the internal energy of the piece of copper. [2]

11 a   Explain what is meant by *specific heat capacity* and *heat capacity*. [3]
    b   Copper has a specific heat capacity of $385\,\mathrm{J\,kg^{-1}\,K^{-1}}$. Calculate the amount heat energy required to raise temperature of $65\,g$ of copper from $25\,°C$ to $65\,°C$. [2]

12 A bicycle and a rider have a total mass of $90\,kg$. The bicycle is travelling along a horizontal road at a constant speed of $6.2\,\mathrm{m\,s^{-1}}$. The brakes are applied until the bicycle and rider comes to rest. During the braking process, 65% of the kinetic energy of the bicycle and rider is converted into thermal energy in the brake blocks. The total mass of the brake blocks is $0.15\,kg$. The specific heat capacity of the brake blocks is $1180\,\mathrm{J\,kg^{-1}\,K^{-1}}$. Calculate:
    a   the initial kinetic energy of the bicycle and rider [2]
    b   the amount of kinetic energy converted into thermal energy in the brake blocks [1]
    c   the maximum temperature rise of the brake blocks. [3]

13 Describe an electrical method to measure the specific heat capacity of water. Include a circuit diagram and explain how the measurements in the experiment are used to determine the specific heat capacity of water. [8]

14 a Define the terms *specific latent of fusion* and
       specific latent heat of vaporisation.   [3]

   b 25 g of ice at −12 °C is added to a polystyrene cup
      containing 180 g of water at 26 °C.
      Calculate:

      i   the amount of thermal energy required to
          convert the ice to water at 0 °C   [3]

      ii the final temperature of the water.   [3]
          [specific heat capacity of water =
          4.2×10^3 J kg K^{-1},
          specific heat capacity of ice = 2.1×10^3 J kg K^{-1},
          specific latent heat of fusion of ice =
          3.3×10^5 J kg^{-1}.]

15 a Explain what is meant by the terms *evaporation*
      and *boiling*.   [2]

   b State one similarity between the processes of
      evaporation and boiling.   [1]

   c State two differences between the processes of
      evaporation and boiling.   [2]

16 Explain the following:

   a Sweating helps an athlete cool down.   [3]

   b The temperature of a liquid drops as evaporation
      takes place.   [3]

17 Describe an experiment to find the specific latent
   heat of fusion of ice.   [6]

18 Describe an experiment to find the specific latent
   heat of vaporisation of water.   [6]

16.1 Ideal gases

Learning outcomes

On completion of this section, you should be able to:

- state the gas laws
- use the equation of state for an ideal gas
- understand the concept of absolute zero.

Definition

The volume of a fixed mass of gas is directly proportional to its absolute temperature, provided that the pressure is kept constant.

Equation

$V \propto T$

$V = kT$

V – volume of gas/m^3
T – absolute temperature/K
k – proportionality constant

Definition

The pressure of a fixed mass of gas is inversely proportional to its volume, provided that the temperature is kept constant.

Equation

$p \propto \dfrac{1}{V}$

$pV = k$

V – volume of gas/m^3
p – pressure of the gas/Pa
k – proportionality constant

Charles' law

A fixed mass of gas is heated at constant pressure. For each temperature measured, a graph of volume against temperature is plotted. Figure 16.1.1 shows the result of such an experiment. When the graph is extrapolated, it cuts the temperature axis at $-273.15\,°C$.

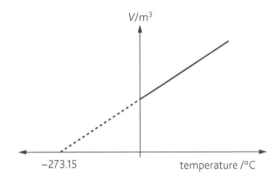

Figure 16.1.1 Charles' law

Charles' law can be used in the form $\dfrac{V_1}{T_1} = \dfrac{V_2}{T_2}$

Boyle's law

A fixed mass of gas is compressed at a constant temperature. For each pressure, the volume of the gas is recorded. The results of this experiment are shown in Figure 16.1.2.

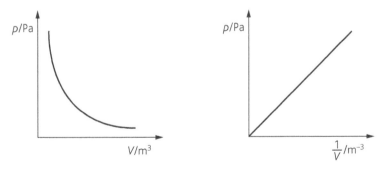

Figure 16.1.2 Boyle's law

Boyle's law can be used in the form $p_1 V_1 = p_2 V_2$.

Pressure law

A fixed mass of gas is heated at constant volume. For each pressure, the temperature of the gas is recorded. The results of this experiment are shown in Figure 16.1.3. When the graph is extrapolated, it cuts the temperature axis at $-273.15\,°C$.

The pressure law can be used in the form $\dfrac{p_1}{T_1} = \dfrac{p_2}{T_2}$.

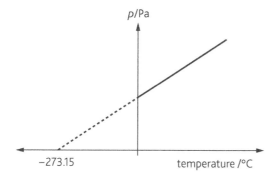

Figure 16.1.3 *The pressure law*

Absolute zero

The experimental data from Charles' law and the Pressure law show that when the graphs are extrapolated, they pass through −273.15°C. This temperature is called **absolute zero**. It is believed to be the lowest temperature possible. It is the zero of the absolute or Kelvin scale of temperature. The SI unit of temperature on this scale is the kelvin (K).

The equation of state

All the gas laws can be combined into one equation known as the equation of state. The constant in this equation is R and is called the molar gas constant. It is equal to $8.31\,\mathrm{J\,mol^{-1}\,K^{-1}}$. An **ideal gas** is one that obeys the gas laws.

Example

An ideal gas is contained in a cylinder which is surrounded by insulation to prevent heat losses. A piston is attached to one end of the cylinder. The initial volume of the gas is $1.3 \times 10^{-4}\,\mathrm{m^3}$. The pressure and temperature of the gas are $1.2 \times 10^5\,\mathrm{Pa}$ and $310\,\mathrm{K}$ respectively.

i Calculate the number of moles of gas in the cylinder.

ii The piston is moved to compress the gas. The volume decreases to $1.9 \times 10^{-5}\,\mathrm{m^3}$. The temperature increases to $650\,\mathrm{K}$. Calculate the new pressure.

i Using the equation of state $\qquad pV = nRT$

The number of moles of gas in the cylinder $n = \dfrac{pV}{RT}$

$$= \frac{1.2 \times 10^5 \times 1.3 \times 10^{-4}}{8.31 \times 310}$$

$$= 6.06 \times 10^{-3}\,\mathrm{mols}$$

ii Using the equation of state $pV = nRT$, n is unchanged.

The new pressure of the gas is $\qquad p = \dfrac{nRT}{V}$

$$= \frac{6.06 \times 10^{-3} \times 8.31 \times 650}{1.9 \times 10^{-5}}$$

$$= 1.72 \times 10^6\,\mathrm{Pa}$$

Definition

The pressure of a fixed mass of gas is directly proportional to its absolute temperature, provided that the volume is kept constant.

Equation

$p \propto T$

$p = kT$

T – absolute temperature/K
p – pressure of the gas/Pa
k – proportionality constant

Equation

$pV = nRT$

n – number of moles of gas
p – pressure/Pa
V – volume/m³
T – absolute temperature/K
R – molar gas constant
$\quad(8.31\,\mathrm{J\,mol^{-1}\,K^{-1}})$

Key points

■ Charles' law states that the volume of a fixed mass of gas is directly proportional to its absolute temperature, provided that the pressure is kept constant.

■ Boyle's law states that the pressure of a fixed mass of gas is inversely proportional to its volume, provided that the temperature is kept constant.

■ The Pressure law states that the pressure of a fixed mass of gas is directly proportional to its absolute temperature, provided that the volume is kept constant.

■ An ideal gas is one that obeys the gas laws.

Assumptions of the kinetic theory of gases

The **kinetic theory of gases** assumes that a gas is made up of many small particles moving randomly at high speeds.

Assumptions of the kinetic theory

- All the molecules that make up the gas are similar (same masses).
- The collisions are elastic (kinetic energy is conserved).
- The intermolecular forces between the molecules of the gas are negligible, except during collisions.
- The volume of the gas molecules is negligible when compared to the volume occupied by the gas.
- The time of collisions is negligible when compared with the time between collisions.
- A large number of molecules are present; therefore the rules of statistical analysis can be applied.
- The motion of the molecules is random.
- Newtonian mechanics apply.

Using the kinetic theory to explain why a gas exerts pressure

Suppose a container is filled with a gas. The gas exerts a pressure on the walls of the container. This pressure can be measured using a manometer or a bourdon gauge. The kinetic theory can be used to explain why the gas exerts a pressure on the inner walls of the container. The molecules of the gas are moving around in a random manner colliding with the walls of the container. When the molecules collide with the walls of the container, they undergo a change in momentum, because their direction changes. According to Newton's second law of motion $\left(F = \dfrac{mv - mu}{\Delta t} = \dfrac{\Delta p}{\Delta t}\right)$, the molecules exert a force on the walls of the container. This force is exerted over the surface area of the inner walls of the container. As a result a pressure is exerted on it $\left(p = \dfrac{F}{A}\right)$.

If the temperature of the gas is increased, the pressure exerted on the walls of the container increases. The kinetic theory can be used to explain why the temperature increases. When the temperature of the gas is increased, the average kinetic energy of the molecules increases. This means that the collisions with the walls of the container occur more frequently. The rate of change of momentum is much greater and therefore the molecules exert a greater force on the walls of the container. A greater force means that the pressure exerted on the inner walls of the container increases.

Mean square speed and root mean square (r.m.s.) speed

Suppose a gas is made up of N atoms having speeds as follows:

$$c_1, c_2, c_3, \ldots, c_N$$

If you were to determine the square of each speed and find the mean value, you would have calculated the **mean square speed** of the atoms

that make up the gas. Mean square speed can be represented as either $\bar{c}^2$ or $\langle c^2 \rangle$.

The **root mean square** (r.m.s.) **speed** is the square root of the mean square speed.

Example

Five gas molecules have speeds of $100\,m\,s^{-1}$, $-250\,m\,s^{-1}$, $300\,m\,s^{-1}$, $275\,m\,s^{-1}$ and $-200\,m\,s^{-1}$.

Calculate

a The mean square speed of the gas molecules.

b The root mean square speed of the gas molecules.

c The mean speed of the molecules.

a $\langle c^2 \rangle = \dfrac{100^2 + (-250)^2 + 300^2 + 275^2 + (-200)^2}{5} = 5.56 \times 10^4\,m\,s^{-1}$

b $\sqrt{\langle c^2 \rangle} = \sqrt{\dfrac{100^2 + (-250)^2 + 300^2 + 275^2 + (-200)^2}{5}} = 236\,m\,s^{-1}$

c $\bar{c} = \dfrac{100 + (-250) + 300 + 275 + (-200)}{5} = 45\,m\,s^{-1}$

Derivation of the pressure exerted by a gas

Consider a cubic container of side l, containing N gas molecules, each of mass m (Figure 16.2.1).

If you were to consider a single molecule, it would be moving about randomly. Suppose at one instant in time its velocity is c (Figure 16.2.2). This velocity can be resolved into three components. (O_x, O_y, O_z). The molecule has velocity components as follows:

In the Ox direction – c_x

In the Oy direction – c_y

In the Oz direction – c_z

Also, $c^2 = c_x^2 + c_y^2 + c_z^2$

Assuming that the molecule that was chosen moves back and forth and strikes the face ABCD.

The distance travelled by the molecule $= 2l$

The time between collisions with the face ABCD $= \dfrac{\text{distance}}{\text{speed}} = \dfrac{2l}{c_x}$

Momentum of molecule just before hitting face ABCD $= mc_x$

Momentum of molecule just after hitting face ABCD $= -mc_x$

Change in momentum $= (-mc_x) - mc_x = -2mc_x$

Rate of change of momentum $= \dfrac{\Delta p}{\Delta t} = \dfrac{2mc_x}{\left(\dfrac{2l}{c_x}\right)} = \dfrac{mc_x^2}{l}$

(The – sign is removed because we are interested in the force acting on the face ABCD).

Therefore, the force exerted on the face ABCD is $\dfrac{mc_x^2}{l}$

Equation

Mean square speed $=$

$\langle c^2 \rangle = \dfrac{c_1^2 + c_2^2 + c_3^2 + \ldots + c_N^2}{N}$

Equation

Root mean square speed $=$

$\sqrt{\langle c^2 \rangle} = \sqrt{\dfrac{c_1^2 + c_2^2 + c_3^2 + \ldots + c_N^2}{N}}$

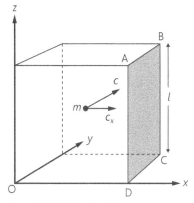

Figure 16.2.1 Deriving the pressure exerted by a gas

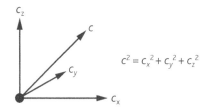

$c^2 = c_x^2 + c_y^2 + c_z^2$

Figure 16.2.2 Resolving the velocity into components along the three axes

The pressure exerted on the face ABCD $= \dfrac{\text{force}}{\text{area}} = \dfrac{\left(\dfrac{mc_x^2}{l}\right)}{l^2} = \dfrac{mc_x^2}{l^3}$

But the volume of the cube $V = l^3$

The pressure exerted on the face ABCD $= \dfrac{\text{force}}{\text{area}} = \dfrac{\left(\dfrac{mc_x^2}{l}\right)}{l^2} = \dfrac{mc_x^2}{V}$

This is the pressure exerted by only one molecule, acting on the face ABCD.

Since there are N molecules, all of them must be considered.

Pressure (exerted by first molecule) $\quad p_1 = \dfrac{mc_{x_1}^2}{V}$

Pressure (exerted by second molecule) $\quad p_2 = \dfrac{mc_{x_2}^2}{V}$

Pressure (exerted by third molecule) $\quad p_3 = \dfrac{mc_{x_3}^2}{V}$

Pressure (exerted by Nth molecule) $\quad p_N = \dfrac{mc_{x_N}^2}{V}$

Total pressure acting of face ABCD $= p_1 + p_2 + p_3 + \ldots + p_N$

$$= \dfrac{mc_{x_1}^2}{V} + \dfrac{mc_{x_2}^2}{V} + \dfrac{mc_{x_3}^2}{V} + \ldots + \dfrac{mc_{x_N}^2}{V}$$

$$= \dfrac{m}{V}\left(c_{x_1}^2 + c_{x_2}^2 + c_{x_3}^2 \ldots + c_{x_N}^2\right)$$

This can be written in terms of mean square speeds.

Let the mean square speed be $\langle c^2 \rangle = \dfrac{c_{x_1}^2 + c_{x_2}^2 + c_{x_3}^2 + \ldots + c_{x_N}^2}{N}$

Therefore $\quad p = \dfrac{Nm\langle c_x^2 \rangle}{V}$

Since the motion of the molecules is random, we could have obtained the equation for p in terms of $\langle c_y^2 \rangle$ or $\langle c_z^2 \rangle$.

So $\quad p = \dfrac{Nm\langle c_x^2 \rangle}{V} = \dfrac{Nm\langle c_y^2 \rangle}{V} = \dfrac{Nm\langle c_z^2 \rangle}{V}$

Therefore, $\quad \langle c_x^2 \rangle = \langle c_y^2 \rangle = \langle c_z^2 \rangle$

But, $\quad c^2 = c_x^2 + c_y^2 + c_z^2$

Therefore, $\quad \langle c^2 \rangle = \langle c_x^2 \rangle + \langle c_y^2 \rangle + \langle c_z^2 \rangle$

Therefore, $\quad \langle c^2 \rangle = 3\langle c_x^2 \rangle$

Or $\quad \langle c_x^2 \rangle = \dfrac{1}{3}\langle c^2 \rangle$

Therefore, the pressure of the gas now becomes $p = \dfrac{1}{3}\dfrac{Nm}{V}\langle c^2 \rangle$

But, density $\rho = \dfrac{\text{mass}}{\text{volume}} = \dfrac{Nm}{V}$

Therefore, the pressure is $\qquad p = \dfrac{1}{3}\rho\langle c^2 \rangle$

Equation

$pV = \dfrac{1}{3}Nm\langle c^2 \rangle$

p – pressure of the gas/Pa
V – volume of gas/m³
N – number of molecules
m – mass of one molecule/kg
$\langle c^2 \rangle$ – mean square speed of the molecules

Total kinetic energy of a monatomic gas

The equation $p = \frac{1}{3}\frac{Nm}{V}\langle c^2 \rangle$ relates a macroscopic property (pressure) to several microscopic properties (number of molecules, mass of the molecules and the speed of the molecules). Let us now compare this equation with the equation of state for an ideal gas ($pV = nRT$).

$$pV = \frac{1}{3}Nm\langle c^2 \rangle \qquad (1)$$

$$pV = nRT \qquad (2)$$

By comparing equations (1) and (2) we obtain the following.

$$\frac{1}{3}Nm\langle c^2 \rangle = nRT$$

Multiply both sides of the equation by $\frac{3}{2}$

$$\frac{3}{2}\left(\frac{1}{3}Nm\langle c^2 \rangle\right) = \frac{3}{2}nRT$$

$$\frac{1}{2}\left(Nm\langle c^2 \rangle\right) = \frac{3}{2}nRT$$

Therefore, the total kinetic energy of all the gas molecules is $\frac{3}{2}nRT$.

Therefore, the kinetic energy of the gas molecules is proportional to the absolute temperature of the gas ($E_k \propto T$).

In order to determine the average kinetic energy of a single molecule, you divide the expression for the total kinetic energy by the number of molecules present.

Total kinetic energy of all the molecules $= \frac{1}{2}\left(Nm\langle c^2 \rangle\right) = \frac{3}{2}nRT$

Kinetic energy of a single molecule $= \frac{1}{2}\left(m\langle c^2 \rangle\right) = \frac{3}{2}\frac{nRT}{N}$

But $N = nN_A$, where N is the total number of molecules, n is the number of moles and N_A is the Avogadro's constant (6.02×10^{23}).

And $k = \frac{R}{N_A}$, where k is known as the Boltzmann constant.

Therefore, the kinetic energy of a single molecule $= \frac{3}{2}kT$

Example

A balloon contains 0.60 mol of helium at 310 K. Calculate:

a the number of helium atoms in the balloon
b the average kinetic energy of a helium atom in the balloon
c the total kinetic energy of the helium atoms in the balloon.
   [Boltzmann constant $= 1.38 \times 10^{-23}$ J K^{-1}
   Avogadro constant $= 6.02 \times 10^{23}$]

a Number of atoms $= N = nN_A = 0.60 \times 6.02 \times 10^{23} = 3.612 \times 10^{23}$.

b Average kinetic energy $= \frac{3}{2}kT = \frac{3}{2} \times 1.38 \times 10^{-23} \times 310$ J
   $= 6.42 \times 10^{-21}$ J

c Total kinetic energy $= 3.612 \times 10^{23} \times 6.42 \times 10^{-21} = 2.32 \times 10^3$ J

Definition

Kinetic energy of all the gas molecules

$= \frac{1}{2}Nm\langle c^2 \rangle = \frac{3}{2}nRT$

m – mass of gas molecule/kg
N – total number of gas molecules
n – number of moles of gas
$\langle c^2 \rangle$ – mean square speed of the molecules
R – molar gas constant (8.31 J mol^{-1} K^{-1})
T – absolute temperature/K

Definition

Translational kinetic energy of a gas molecule

$\frac{1}{2}m\langle c^2 \rangle = \frac{3}{2}kT$

m – mass of gas molecule/kg
$\langle c^2 \rangle$ – mean square speed of the molecules
k – Boltzmann constant (1.38 × 10^{-23} J K^{-1})
T – absolute temperature/K

Key points

■ The kinetic theory has several assumptions.

■ The kinetic theory can be used to explain why a gas exerts pressure.

■ The average translational kinetic energy of a monatomic gas molecule is given by $\frac{3}{2}kT$

■ The total kinetic energy of all the molecules in a gas is given by $\frac{3}{2}nRT$

17.1 The first law of thermodynamics

Learning outcomes

On completion of this section, you should be able to:

- state the first law of thermodynamics

- derive an expression for the work done by a gas.

Definition

The first law of thermodynamics states that the change in internal energy of a system is equal to the energy supplied to the system plus the work done on the system.

The first law of thermodynamics

The **first law of thermodynamics** is simply the principle of conservation of energy. It states that the change in internal energy of a system is equal to the energy supplied to the system plus the work done on the system.

The first law of thermodynamics can be expressed mathematically as follows:

Equation

The first law of thermodynamics

$$\Delta U = \Delta Q + \Delta W$$

ΔU – change in internal energy of the system/J
ΔQ – energy supplied to the system/J
ΔW – work done on the system/J

It is important to understand the sign convention used when applying the equation above.

ΔU is positive (+) when the internal energy is increasing.

ΔU is negative (−) when the internal energy is decreasing.

ΔQ is positive (+) when energy is being supplied to the system.

ΔQ is negative (−) when energy is being removed from the system.

ΔW is positive (+) when work is being done **on** the system.

ΔW is negative (−) when work is being done **by** the system.

The work done by a gas

Consider a gas cylinder that has a movable piston attached to one end (Figure 17.1.1). The cross-sectional area of the piston is A. The piston traps the gas inside the cylinder. As long as there is a pressure difference between the gas contained in the cylinder and the external atmospheric pressure, the piston will begin moving freely. Now, suppose a gas of volume V is inside the cylinder. The piston will move to the right until the pressure of the gas is equal to atmospheric pressure p.

If the valve is opened to allow more gas to enter the cylinder, the pressure inside the cylinder will begin to increase (Figure 17.1.2). It increases because more molecules of gas are being added and there are going to be more collisions per second acting on the inner walls of the cylinder. The piston will begin to move to the right. The gas slowly pushes back the atmosphere with a force F. The piston stops moving when the internal and external pressures are the same. The piston moves through a distance x.

The work done by the gas against atmospheric pressure is given by

$$\Delta W = \text{force} \times \text{distance moved in the direction of the force}$$

$$= F \times \Delta x$$

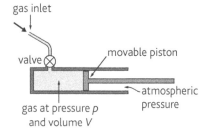

Figure 17.1.1

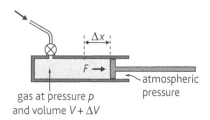

Figure 17.1.2

But pressure $= \dfrac{\text{force}}{\text{area}}$

$\therefore \qquad F = p \times A$

Work done by the gas against atmospheric pressure $= p \times A \times \Delta x$

But the change in volume $\Delta V = A \times \Delta x$

Work done by the gas against atmospheric pressure $\Delta W = p\Delta V$

When work is being done by the gas, ΔW is negative.

The same expression can be used when the gas is being compressed.

In this case, ΔW is positive.

Equation

$\Delta W = p\Delta V$

ΔW – work done /J
p – pressure/Pa
ΔV – change in volume/m³

Applying the first law of thermodynamics

Example

A fixed mass of an ideal gas absorbs 1500 J of heat energy and expands under a constant pressure of 2.5×10^4 Pa from a volume of 2.1×10^{-2} m³ to a volume of 4.2×10^{-2} m³. Calculate the change in internal energy of the gas.

Heat supplied to the gas is $\Delta Q = 1500$ J

Since the pressure remains constant,

Work done by gas $\Delta W = p\Delta V$

$\qquad = 2.5 \times 10^4 (4.2 \times 10^{-2} - 2.1 \times 10^{-2})$

$\qquad = 420$ J

Using the first law of thermodynamics

Change in internal energy of gas $\Delta U = \Delta Q + \Delta W$

$\qquad = 1500 + (-420)$

$\qquad = 1080$ J

Note that since work was done by the gas a minus sign is needed in front of ΔW.

Example

An ideal gas is contained in a cylinder which is surrounded by insulation to prevent heat losses. The initial volume of the gas is 2.5×10^{-4} m³. The pressure and temperature of the gas is 1.1×10^5 Pa and 320 K respectively.

i Calculate the number of moles of gas in the cylinder.

ii The piston is moved to compress the gas. The volume decreases to 2.1×10^{-5} m³. The temperature increases to 750 K. Calculate the new pressure of the gas and explain why the temperature of the gas increases.

ii The work done on the gas is 86 J. Determine the increase in the internal energy of the gas.

i Using the equation of state $\qquad pV = nRT$

The number of moles of gas in the cylinder $n = \dfrac{pV}{RT}$

$\qquad = \dfrac{1.1 \times 10^5 \times 2.5 \times 10^{-4}}{8.31 \times 320}$

$\qquad = 1.03 \times 10^{-2}$ mol

ii Using the equation of state, $pV = nRT$ is unchanged.

The new pressure of the gas is $\quad p = \dfrac{nRT}{V}$

$$= \frac{1.03 \times 10^{-2} \times 8.31 \times 750}{2.1 \times 10^{-5}}$$

$$= 3.06 \times 10^{6}\,\text{Pa}$$

The molecules of the gas are moving about randomly at high velocities. When the molecules collide with the moving piston, they rebound with a greater velocity. This means that the average kinetic energy of all the molecules increase. Since the temperature of a gas is proportional to the mean kinetic energy of all its molecules, there is an increase in temperature of the compressed gas.

iii $\Delta W = +86\,\text{J}$

Energy supplied to the system $\Delta Q = 0\,\text{J}$ since the cylinder is insulated.

Using the first law of thermodynamics

The change in internal energy of the gas $\Delta U = \Delta Q + \Delta W$

$$= 0 + 86$$

$$= +86\,\text{J}$$

Example

A cylinder is fitted with a piston which can move without friction contains 0.055 mols of an ideal gas at a temperature of 300 K and a pressure of 1×10^5 Pa (Figure 17.1.3).

Calculate:

i the volume of the gas

ii the internal energy of the gas.

Suppose the temperature of the gas is increased to 360 K and the pressure is kept constant.

Calculate:

iii the change in internal energy of the gas

iv the external work done by the gas

v the total amount of energy supplied to the gas.

[molar gas constant $= 8.31\,\text{J}\,\text{mol}^{-1}\,\text{K}^{-1}$]

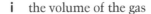

i Using the equation of state $pV = nRT$

Volume of gas $V = \dfrac{nRT}{p} = \dfrac{0.055 \times 8.31 \times 300}{1 \times 10^5} = 1.37 \times 10^{-3}\,\text{m}^3$

ii The internal energy of an ideal gas is dependent on the temperature of the gas. The temperature of the gas is dependent on the kinetic energy of all the molecules that make up the gas. The total kinetic energy of an ideal gas is given by

$$E_{\text{K}} = \frac{3}{2}nRT$$

$\therefore \qquad U = \dfrac{3}{2}nRT$

Internal energy $U = \dfrac{3}{2}nRT = \dfrac{3}{2} \times 0.055 \times 8.31 \times 300 = 206\,\text{J}$

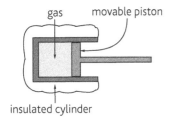

gas   movable piston

insulated cylinder

Figure 17.1.3

iii Change in temperature of the gas $= 360 - 300 = 60\,\mathrm{K}$

Change in internal energy $\Delta U = \dfrac{3}{2}nR\Delta T$

$$= \dfrac{3}{2} \times 0.055 \times 8.31 \times 60 = 41\,\mathrm{J}$$

iv External work done by the gas is given by $\Delta W = p\Delta V$

The new volume of the gas must first be determined.

$$V = \dfrac{nRT}{p} = \dfrac{0.055 \times 8.31 \times 360}{1 \times 10^5} = 1.65 \times 10^{-3}\,\mathrm{m^3}$$

External work done by the gas $\Delta W = p\Delta V$

$$= 1 \times 10^5(1.65 \times 10^{-3} - 1.37 \times 10^{-3}) = 28\,\mathrm{J}$$

v Using the first law of thermodynamics $\Delta U = \Delta Q + \Delta W$

Since work was done by the gas $\Delta W = -28\,\mathrm{J}$

Total energy supplied to the gas $\Delta Q = \Delta U - \Delta W = 41 - (-28) = 69\,\mathrm{J}$

Example

At a temperature of $100\,^{\circ}\mathrm{C}$ and pressure of $1.01 \times 10^5\,\mathrm{Pa}$, 1 kg of steam occupies $1.67\,\mathrm{m^3}$. At the same temperature and pressure, 1 kg of water occupies $1.04 \times 10^{-3}\,\mathrm{m^3}$.

If 1 kg of water at $100\,^{\circ}\mathrm{C}$ is converted to 1 kg of steam at $100\,^{\circ}\mathrm{C}$.

Calculate:

i the energy supplied to produce this change

ii the work done against the atmosphere

iii the increase in internal energy

iv the increase in potential energy of the molecules of water.

$[l_v = 2.26 \times 10^6\,\mathrm{J\,kg^{-1}}]$

i Energy supplied to produce change $Q = ml_v$

$$= 1 \times 2.26 \times 10^6$$

$$= 2.26 \times 10^6\,\mathrm{J}$$

ii Work done against the atmosphere

$$\Delta W = p\Delta V$$

$$= 1.01 \times 10^5(1.67 - 1.04 \times 10^3)$$

$$= 1.69 \times 10^5\,\mathrm{J}$$

iii Increase in internal energy $\Delta U = \Delta Q + \Delta W$

$$= 2.26 \times 10^6 + (-1.69 \times 10^5)$$

$$= 2.09 \times 10^6\,\mathrm{J}$$

iv **At a change of state, there is no change in temperature. All the energy supplied is used to increase the potential energy of the molecules and not their kinetic energies.**

Therefore, the increase in potential energy of the molecules $= 2.09 \times 10^6\,\mathrm{J}$.

Key points

- The first law of thermodynamics states the change in internal energy of a system is equal to the energy supplied to it plus the work done on the system.

- The work done by a gas at constant pressure is equal to the pressure multiplied by the change in volume of the gas.

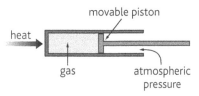

Figure 17.2.1 *Heating a gas at constant pressure*

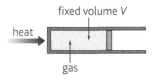

Figure 17.2.2 *Heating a gas at constant volume*

Equations

$E_H = nC_p\Delta\theta$

$E_H = nC_v\Delta\theta$

E_H – energy

n – number of moles

C_p – molar heat capacity at constant pressure

C_v – molar heat capacity at constant volume

$\Delta\theta$ – change in temperature of the gas

Molar heat capacities

The **molar heat capacity** of a substance is the amount of thermal energy required to increase the temperature of one mole of it by one degree. In the case of a gas, large changes in pressure and volume occur when supplied with thermal energy. The change in volume and pressure when a solid or liquid is heated is negligible. The heat capacity of a gas can be defined under two particular conditions.

- Molar heat capacity at constant pressure.
- Molar heat capacity at constant volume.

The **molar heat capacity of a gas at constant pressure** C_p, is the amount of energy required to raise the temperature of one mole of a gas by one degree, when the pressure remains constant.

The **molar heat capacity at constant volume** C_v, is the amount of energy required to raise the temperature of one mole of a gas by one degree when the volume remains constant.

The molar heat capacity at constant pressure is greater than the molar heat capacity at constant volume.

Consider a gas contained in a cylinder at a pressure p. Heat energy is then supplied to the gas. In order for the pressure to remain constant, the gas must expand and move the piston. In this process, the gas has to do external work by moving the piston. Also, the heat supplied increases the potential and kinetic energy of the gas molecules (Figure 17.2.1).

Now consider the same gas contained in a cylinder having a volume V. When the gas is heated, all the energy supplied to it is used to increase its temperature. Since, the volume of the container is fixed, the gas does no external work. This means that less energy is required to raise the temperature of a gas when its volume is fixed. It follows that $C_p > C_v$.

Molar heat capacity at constant pressure and constant volume are related as follows (Figure 17.2.2):

$C_p - C_v = R$, where R is the molar gas constant ($R = 8.31\,\mathrm{J\,mol^{-1}\,K^{-1}}$).

Using *p–V* diagrams

An indicator diagram is a graph showing how the pressure p of a gas varies with its volume V during a change. The work done in each stage can be determined. Suppose a gas has a pressure and volume, such that its state is represented by the point A. The gas then expands while its pressure remains fixed. Its new state is represented by the point B. The work done by the gas is given by $\Delta W = p\Delta V$. If the volume of the gas is kept fixed and its pressure is increased, the new state would be represented by the point C. No work is done by the gas during this change. Suppose the gas is now compressed, while keeping the pressure fixed. The new state would be represented by the point D. The work done on the gas is given by $\Delta W = p\Delta V$. If the volume of the gas is kept fixed and its pressure is decreased, the new state would be represented by the point A. No work is done by the gas during this change. During the entire cycle, the change in internal energy of the gas is zero. The internal energy of a gas is dependent on its state. The gas starts off at the point A and returns to the point A, meaning that it hasn't changed state. The net effect of all the changes is zero.

Example

A fixed mass of an ideal gas in a heat pump undergoes a cycle of changes of pressure, volume and temperature as shown in Figure 17.2.3 (not drawn to scale).

No heat is supplied to the gas from A→B and C→D. The increase in internal energy of the gas is as follows

A→B 1300 J   B→C −1200 J   C→D −450 J

Using the first law of thermodynamics and the data supplied, determine the following:

i   the work done on the gas from A to B and C to D
ii the work done on the gas from B to C and D to A
iii the heat supplied to the gas from B to C
iv the increase in internal energy of the gas from D to A
v   the heat supplied to the gas from D to A
vi the number of moles of gas being used.

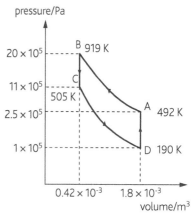

Figure 17.2.3

First law of thermodynamics $\Delta U = \Delta Q + \Delta W$

i From A to B, $\Delta W = \Delta U + \Delta Q = 1300 - 0 = 1300\,\text{J}$
   From C to D, $\Delta W = \Delta U + \Delta Q = -450 - 0 = -450\,\text{J}$

ii From B to C there is no change in volume of the gas, therefore $\Delta V = 0$. This means that no work is done on or by the gas.
   ∴ $\Delta W = 0$

   From D to A there is no change in volume of the gas, therefore $\Delta V = 0$. This means that no work is done on or by the gas.
   ∴ $\Delta W = 0$

iii Heat supplied to the gas for B to C, $\Delta Q = \Delta U - \Delta W$
                    $= 1200 - 0 = 1200\,\text{J}$

iv The internal energy of a gas is dependent on its state. The internal energy is dependent on temperature. Since the temperature at A is fixed, the total change in internal energy of the gas during the complete cycle is zero.
   From A to B, $\Delta U = +1300\,\text{J}$
   From B to C, $\Delta U = -1200\,\text{J}$
   From C to D, $\Delta U = -450\,\text{J}$
   From D to A, $\Delta U = ?$

   All these internal energy values should add up to zero, because ΔU from A to A is zero.
   From D to A, $\Delta U = 0 - (1300 + -1200 + -450) = +350\,\text{J}$

v Heat supplied to the gas for D to A,
   $\Delta Q = \Delta U - \Delta W = 350 - 0 = 350\,\text{J}$

vi Using the equation of state $pV = nRT$
   The number of moles of gas in the cylinder $n = \dfrac{pV}{RT}$

$$= \frac{2.5 \times 10^5 \times 1.8 \times 10^{-3}}{8.31 \times 492} = 0.11\,\text{mols}$$

Key points

- The molar heat capacity of a gas at constant pressure C_p, is the amount of energy required to raise the temperature of one mole of a gas by one degree, when the pressure remains constant.

- The molar heat capacity at constant volume C_v, is the amount of energy required to raise the temperature of one mole of a gas by one degree when the volume remains constant.

- The molar heat capacity at constant pressure is greater than the molar heat capacity at constant volume.

- p–V diagrams can be used to illustrate the changes in pressure and volume of a gas.

Revision questions 8

Answers to questions that require calculation can be found on the accompanying CD.

1 Explain what is meant by:
  a an ideal gas [1]
  b absolute zero. [1]

2 A bubble of gas rises from the bottom of a pond to the surface. The pressure at the bottom of the pond is 2.5×10^5 Pa and the pressure at the surface is 1.0×10^5 Pa. The volume of the bubble at the bottom of the pond is 2.5 cm³. Calculate the volume of the bubble at the surface. [3]

3 A cylinder contains 1.8 mol of a gas at room temperature, 25 °C and the pressure inside the cylinder is 3.6×10^5 Pa. The temperature increases to 60 °C when an additional 4.5 mol of gas is pumped into the cylinder. Calculate the new pressure inside the cylinder. [5]

4 The volume of a gas is V at a temperature of 220 K. The gas is heated to 330 K at constant pressure. What is the percentage increase in the volume of the gas? [3]

5 a Explain what is meant by an ideal gas. [1]
  b Write down the equation of state for an ideal gas, explaining the symbols used. [3]
  c An ideal gas has a volume of 1.9×10^{-3} m³ and is at a temperature of 300 K. The pressure of the gas is 1.0×10^5 Pa. The gas is compressed until its volume and pressure is 1.1×10^{-3} m³ and 1.6×10^5 Pa respectively. Calculate:
    i   the number of moles of gas present [2]
    ii the final temperature of the gas. [3]

6 A cylinder of volume 6.5×10^3 m³ contains helium gas at a pressure of 1.8×10^7 Pa and a temperature of 300 K.
  Calculate:
  a the amount of gas in the container [2]
  b the mean kinetic energy of the helium atoms [2]
  c the total kinetic energy of the helium atoms in the container. [3]

7 A fixed mass of an ideal gas absorbs 1100 J of heat energy and expands under a constant pressure of 2.1×10^4 Pa from a volume of 1.9×10^{-2} m³ to a volume of 3.6×10^{-2} m³.
  Calculate the change in internal energy of the gas. [5]

8 A cylinder is fitted with a piston which can move without friction and contains 0.022 mols of an ideal gas at a temperature of 280 K and a pressure of 1.6×10^5 Pa.
  Calculate:
    i   the volume of the gas [2]
    ii the internal energy of the gas. [3]
  Suppose the temperature of the gas is increased to 330 K and the pressure is kept constant.
  Calculate:
    iii the change in internal energy of the gas [2]
    iv   the external work done by the gas [2]
    v   the total amount of energy supplied to the gas. [2]
    [Molar gas constant = 8.31 J mol⁻¹ K⁻¹]

9 a Explain what is meant by:
    i   the molar heat capacity at constant volume
    ii the molar heat capacity at constant pressure. [2]
  b Explain why the molar heat capacity at constant pressure is greater than the molar heat capacity at constant volume for a gas. [3]
  c State the relationship between the two quantities. [1]

10 a State four assumptions of the kinetic theory. [4]
   b The kinetic theory of gases can show that the pressure p and the volume V of an ideal gas is given by the expression
   $$pV = \frac{1}{3} Nm\langle c^2 \rangle$$
   where m is the mass of a gas molecule.
     i   State what the following symbols mean: N and $\langle c^2 \rangle$ [2]
     ii Use the expression to deduce the mean kinetic energy of a gas molecule at a temperature T. [3]
   c i   State what is meant by the internal energy of a substance. [2]
     ii Using the expression derived in **b ii** to explain why the change in internal energy of an ideal gas is proportional to the change in temperature of the gas. [2]

11 A fixed mass of an ideal gas undergoes a cycle ABCA as shown below.

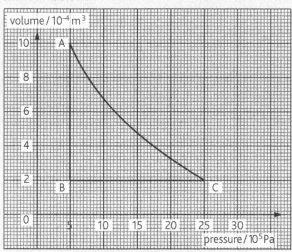

a State the change in internal energy of the gas when it undergoes the cycle ABCA. [1]

b Calculate the work done on the gas during the change A to B. [2]

c Copy and complete the table below relating the energy changes during the cycle ABCA. [6]

Change	Work done on gas/J	Energy supplied to gas/J	Change in internal energy of the gas/J
A → B		-520	
B → C		+650	
C → A		+315	

18.1 Thermal conduction and convection

Thermal energy can be transferred from one point to another by either one or more of the following three methods:

- conduction – a medium is required.
- convection – a medium is required.
- radiation – no medium is required. Energy can be transferred in a vacuum.

Thermal conduction

Thermal conduction is a process by which thermal energy flows from a region of high temperature to a region of low temperature without any net movement of the material itself. Thermal conduction occurs because of one of two mechanisms. These mechanisms are called **lattice vibrations** and **electron diffusion**.

Lattice vibrations

In a non-metal such as ceramic there are no free electrons. If some hot coffee is poured into a ceramic mug, the outside of it becomes hot. In the ceramic mug, adjacent atoms are bonded together by strong bonds. When the coffee is in contact with the inner walls of the ceramic mug, the atoms in it begin to vibrate as they gain energy. As they begin to vibrate, adjacent atoms begin vibrating as well. These atoms then collide with their neighbours and the process repeats until heat flows from the inner walls of the mug to the outside walls of the mug. This process by which thermal energy flows is called **lattice vibrations** (Figure 18.1.1).

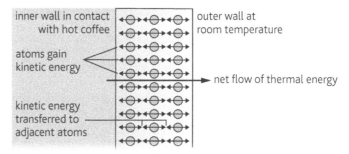

Figure 18.1.1 Conduction by lattice vibrations

Electron diffusion

Metals are very good conductors of heat. If a metal spoon is placed in a pot of hot soup, thermal energy is quickly transmitted up the handle of the spoon. In a metal, there are free electrons moving around randomly within the structure. When the spoon is placed in the pot of hot soup, the atoms in the metal begin vibrating. The free electrons gain kinetic energy every time they collide with a vibrating atom. Since the electrons are moving within the structure, they collide with other atoms and transfer some of their kinetic energy. In this process, thermal energy is being transferred by the fast-moving electrons. This process is called electron diffusion. In this process, thermal energy transfers at a much faster rate than lattice

vibrations. In a metal, some of the thermal energy is transferred by lattice vibrations, but this occurs to a lesser degree than electron diffusion. Metals are good conductors of heat and electricity (Figure 18.1.2).

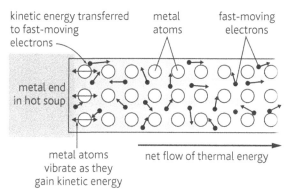

Figure 18.1.2 Conduction by electron diffusion

Thermal conductivity

Conduction occurs at different rates in different thermal conductors. The rate at which thermal energy is transferred by thermal conduction depends on:

- the cross-sectional area
- the material from which the conductor is made
- the temperature gradient across the conductor.

Consider a slab of material of cross-sectional area A and length l. The temperatures at the ends of the conductor are maintained at θ_1 and θ_2 $(\theta_1 > \theta_2)$. This is referred to as steady state conditions.

The sides of the conductor are completely lagged so that there are no heat losses. The temperature gradient is defined as $\frac{\theta_1 - \theta_2}{l}$.

The rate of flow of heat through the conductor $\left(\frac{Q}{t}\right)$ is expressed as follows:

$$\frac{Q}{t} \propto A\frac{\theta_1 - \theta_2}{l}$$

The equation can be re-written as follows:

$$\frac{Q}{t} = -kA\left(\frac{\theta_1 - \theta_2}{l}\right)$$

The constant of proportionality k, is called the coefficient of **thermal conductivity**. The negative sign indicates that thermal energy is being transferred in the direction in which the temperature is decreasing (Figure 18.1.3).

Good thermal conductors have a high coefficient of thermal conductivities.

(e.g. copper = $390\,W\,m^{-1}\,K^{-1}$, silver = $420\,W\,m^{-1}\,K^{-1}$)

Poor thermal conductors have a low coefficient of thermal conductivity.

(e.g. glass = $0.8\,W\,m^{-1}\,K^{-1}$, wood = $0.15\,W\,m^{-1}\,K^{-1}$)

The graph in Figure 18.1.4 shows how the temperature across a conductor varies with distance along it when the sides are insulated. The temperature gradient is constant.

Definition

The coefficient of thermal conductivity is the rate of flow of heat per unit area per unit temperature gradient, when the heat flow is at right angles to the faces of a thin parallel-sided slab of the material, under steady state conditions.

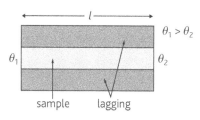

Figure 18.1.3 Coefficient of thermal conductivity

Equation

$$\frac{Q}{t} = -kA\left(\frac{\theta_1 - \theta_2}{l}\right)$$

$\frac{Q}{t}$ – rate of flow of heat/W or $J\,s^{-1}$

k – coefficient of thermal conductivity/$W\,m^{-1}\,K^{-1}$

A – cross-sectional area/m^2

$\frac{\theta_1 - \theta_2}{l}$ – temperature gradient/$K\,m^{-1}$

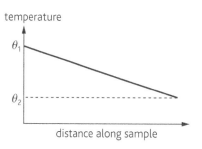

Figure 18.1.4 Temperature difference across a thermal conductor (lagging)

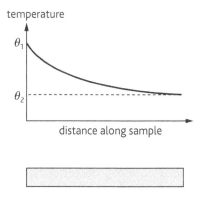

temperature

distance along sample

sample without lagging

Figure 18.1.5 *Temperature difference across a thermal conductor (without lagging)*

The graph in Figure 18.1.5 shows how the temperature across a conductor varies with distance along it when the sides are not insulated. The temperature gradient is not constant. Heat escapes from the sides and the temperature gradient decreases as the distance from the hotter side increases.

Example

A copper rod of length $0.6\,m$ and cross-sectional area $7.85 \times 10^{-5}\,m^2$ is arranged such that it is completely lagged on the sides. One end is maintained at $100\,°C$ and the other end is maintained at $0\,°C$ (Figure 18.1.6). Calculate the rate of flow of thermal energy through the copper rod (thermal conductivity of copper $= 390\,W\,m^{-1}\,K^{-1}$).

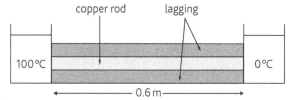

copper rod   lagging

$100\,°C$ $0\,°C$

$0.6\,m$

Figure 18.1.6

Under steady state conditions

$$\frac{Q}{t} = -kA\left(\frac{\theta_1 - \theta_2}{l}\right)$$

$$\frac{Q}{t} = -390 \times 7.85 \times 10^{-5} \times \left(\frac{100 - 0}{0.6}\right)W = 5.10\,W$$

Example

An ideally lagged compound bar consists of a copper bar $12\,cm$ long joined to an aluminium bar of length $20\,cm$ long and of equal cross-sectional area. The free end of the copper bar is maintained at $100\,°C$ and the free end of the aluminium is maintained at $0\,°C$ (Figure 18.1.7). Calculate the temperature at the point at which the copper and aluminium bars are joined.

(Thermal conductivity of copper $= 390\,W\,m^{-1}\,K^{-1}$, thermal conductivity of aluminium $= 220\,W\,m^{-1}\,K^{-1}$)

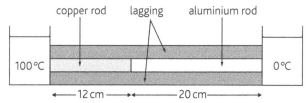

copper rod   lagging   aluminium rod

$100\,°C$ $0\,°C$

$12\,cm$ $20\,cm$

Figure 18.1.7

Under steady state conditions, the rate of flow of heat through the copper and aluminium are the same. (Lagging prevents heat loss from the sides.)

Let the temperature at the point where the two bars meet be θ.

$$\frac{Q}{t} = 390 \times A \times \left(\frac{100 - \theta}{0.12}\right) = 210 \times A \times \left(\frac{\theta - 0}{0.20}\right)$$

$$390 \times \left(\frac{100 - \theta}{0.12}\right) = \left(\frac{210\theta}{0.20}\right)$$

$$25.2\theta = 7800 - 78\theta$$

$$25.2\theta + 78\theta = 7800$$

$$\theta = \frac{7800}{103.2} = 75.6\,°C$$

Convection

In fluids (liquids and gases) the main form of thermal energy transfer is **convection**. Thermal energy is transferred from a region of high temperature to a region of low temperature due to the bulk movement of the fluid as a result of a density change. When some water is heated in a beaker, the liquid at the bottom becomes warmer. It expands and therefore becomes less dense than the liquid above it. The cooler denser liquid from above sinks to the bottom and causes the less dense liquid to rise. The process sets up convection currents (Figure 18.1.8). In this process heat is transferred because of the change in density of the liquid.

Land and sea breezes

The specific heat capacity of water is much larger than that of soil. During the day, the land heats up much faster than the sea. The hot air above the land rises and the cooler air from above the sea rushes in to replace it. This creates a sea breeze (Figure 18.1.9).

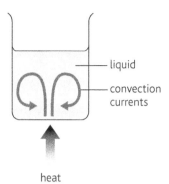

Figure 18.1.8 *Convection currents in a liquid*

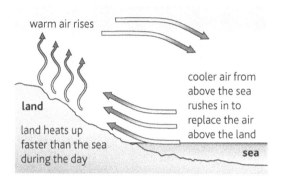

Figure 18.1.9 *The formation of sea breezes (day time)*

During the night, the reverse process occurs. The land cools more quickly than the sea. The air above the sea is warmer and therefore rises. The cooler, denser air from above the land rushes in to replace it. This creates a land breeze (Figure 18.1.10).

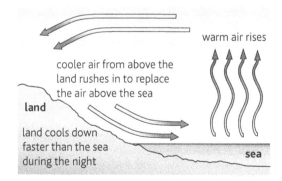

Figure 18.1.10 *The formation of land breezes (night time)*

Ocean currents

Ocean currents occur as a result of the change in density of sea water. As the sun heats up the ocean, warm water rises because its density decreases. Colder denser water rushes in and convection currents are set up.

Key points

- Thermal energy is transferred by conduction, convection and radiation.

- Thermal conduction occurs by electron diffusion or lattice vibrations.

- Electron diffusion is the main mechanism for thermal energy transfer in metals.

- The coefficient of thermal conductivity is the rate of flow of heat per unit area per unit temperature gradient, when the heat flow is at right angles to the faces of a thin parallel-sided slab of the material, under steady state conditions.

- Thermal convection is the transfer of thermal energy from a region of high temperature to a region of low temperature due to the bulk movement of the fluid as a result of a density change.

Learning outcomes

On completion of this section, you should be able to:

- describe an experiment to find the thermal conductivity of a good conductor

- describe an experiment to find the thermal conductivity of a poor conductor.

Experiment to find the thermal conductivity of a good conductor

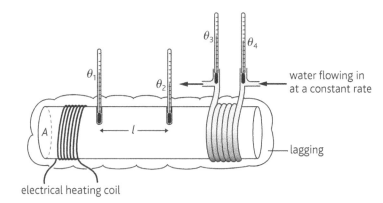

Figure 18.2.1 *Experiment to measure the thermal conductivity of a good conductor*

In the case of good conductors, the length of the bar should be sufficiently long to give a measurable temperature difference. This ensures that there is a measurable temperature gradient. In this experiment, the bar is lagged to prevent heat losses from the sides of the bar. The experiment used to determine the thermal conductivity is shown in Figure 18.2.1. One end of the bar is heated with an electric heater. The other end of the bar is cooled by water flowing through conducting coils at a constant rate. When steady state conditions are reached (i.e the reading on all the thermometers have become steady), θ_1, θ_2, θ_3 and θ_4 are recorded. The rate of flow of heat through the conductor is given by:

$$\frac{Q}{t} = -kA\left(\frac{\theta_1 - \theta_2}{l}\right) \qquad (1)$$

Where A is the cross-sectional area of the conductor and l is the distance between θ_1 and θ_2 and k is the thermal conductivity of the conductor.

The rate at which thermal energy is being removed by the water is given by:

$$\frac{Q}{t} = \frac{m}{t} \times c_w \times (\theta_3 - \theta_4) \qquad (2)$$

where m/t is the rate of flow of the water and c_w is the specific heat capacity of water.

Equation (1) and equation (2) are equated to determine the value of k.

$$-kA\left(\frac{\theta_1 - \theta_2}{l}\right) = \frac{m}{t}c_w(\theta_3 - \theta_4)$$

Experiment to find the thermal conductivity of a poor conductor

In the case of a poor conductor, the sample must be thin and should have a large cross-sectional area. This allows for a large rate of flow of heat through the sample.

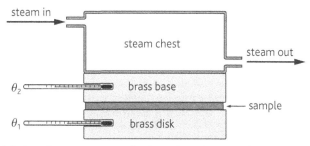

Figure 18.2.2 *Experiment to measure the thermal conductivity of a poor conductor*

In this experiment, the sample is placed between two brass slabs. Thermometers are inserted in holes in the brass slabs. The upper slab is heated using a steam chest (Figure 18.2.2). Since the sample is thin and it is a poor conductor, it can be assumed that heat losses from the sides are negligible. Under steady state conditions, the temperature θ_1 and θ_2 are constant and the rate of flow of heat through the sample is equal to the rate of loss of heat from the bottom of the brass disc. The main form of thermal energy loss is by convection.

The rate of flow of heat through the sample is given by:

$$\frac{Q}{t} = -kA\left(\frac{\theta_1 - \theta_2}{l}\right) \qquad (1)$$

where

$A = \pi r^2$ and r is the radius of the disc and l is the thickness of the sample.

The rate of loss of heat from the bottom brass slab is given by:

$$\frac{Q}{t} = mc \times \left(\frac{\Delta\theta}{\Delta t}\right) \qquad (2)$$

where m is the mass of the brass disc and c is the specific heat capacity of brass.

In order to determine the rate at which the brass disc cools, the brass disc is heated directly with the steam chamber. When the temperature of the brass disc is steady, the sample and steam chamber are removed and the brass disc is covered with an insulator (Figure 18.2.3). The temperature θ_1 is recorded over a period of time until it approximately 5 °C less than the previous steady state value. A cooling curve is then plotted. The slope of the graph at θ_1 is determined (Figure 18.2.4).

Equation (1) and equation (2) are equated to determine the value of k.

$$-kA\left(\frac{\theta_1 - \theta_2}{l}\right) = mc\frac{\Delta\theta}{\Delta t}$$

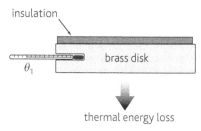

Figure 18.2.3

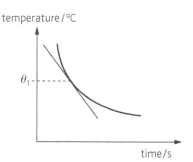

Figure 18.2.4 *Measuring the rate of heat loss by convection from the brass disc*

Key points

- When determining the thermal conductivity of a good conductor, the conductor should be long and thin.

- When determining the thermal conductivity of a poor conductor, the sample should be thin and have a large cross-sectional area.

18.3 Radiation

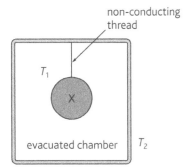

Figure 18.3.1

Thermal radiation

Hot objects emit energy by a process called thermal radiation. The higher the temperature of the object, the more energy is transmitted away per second by thermal radiation. Thermal radiation is made up of **electromagnetic waves** with a continuous range of wavelengths. These wavelengths span from the infrared to the visible regions of the electromagnetic spectrum. When thermal radiation is incident on the surface of a body it can be either reflected or absorbed.

Suppose a small object X, at a temperature T_1 is placed inside an evacuated chamber whose walls are maintained at a temperature T_2 (Figure 18.3.1). Since the chamber is evacuated, heat cannot be transmitted by convection or conduction. All the energy is transmitted by thermal radiation. If $T_1 > T_2$, the object loses energy by thermal radiation until its temperature becomes T_2. If $T_1 < T_2$, the object gains energy by thermal radiation until its temperature becomes T_2. In either case, the temperature will become T_2 eventually. It should be noted that energy absorption and emission does not cease at this point.

Prevost's theory of exchanges states that when a body is at the same temperature as its surroundings its rate of emission of radiation to the surroundings is equal to the rate of absorption of radiation from the surroundings.

Therefore, in the previous demonstration even though the object X is at the same temperature as its surroundings, it is still absorbing and emitting energy by thermal radiation. It is said to be in **thermal equilibrium** with its surroundings.

The rate at which energy flows from an object by thermal radiation depends on the temperature of the object.

It follows that a body which is a good absorber of thermal radiation is also a good emitter of thermal radiation; otherwise the temperature would rise above that of its surroundings.

Good absorbers are also good emitters of thermal radiation.

Suppose equal volumes of boiling water at 100 °C are placed into two identical metal cans, A and B. The outer surface of can A is painted silver and the outer surface of can B is painted black. If the temperature of the water in both cans is measured over a period of time, it can be shown that the water in can B cools faster than that in can A. The black surface radiates heat better than the silvered surface.

The experiment can be repeated using water at room temperature. Equal volumes of water are placed in cans A and B. The two cans are placed at equal distance from a lighted Bunsen burner. A thermometer is used to measure the temperature of water over a period of time. It can be shown that the temperature of the water in can B increases more quickly than the water in can A.

Another example is the drying of crops. In some countries, crops are dried on bitumen surfaces. The black surface absorbs energy by thermal radiation and becomes hot. This is an effective means of drying crops.

A perfect absorber of thermal radiation is called a **black body**. A black body is one that absorbs all the radiation which is incident on it. It should be pointed out that the concept of a black body is a theoretical one.

Energy distribution in the spectrum of a black body radiator

A black body radiator is one that emits thermal radiation which is characteristic of its temperature. Black body radiation from a source at constant temperature consists of a continuous range of wavelengths. Figure 18.3.2 shows a typical spectrum obtained from a black body radiator at a constant temperature.

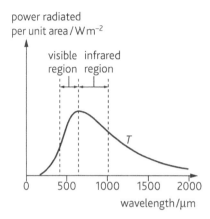

Figure 18.3.2 *Spectrum of a black body radiator*

Since the graph is curved it means that the energy carried away by radiation is not evenly distributed across the range of wavelengths. Also, the area under the graph gives the total energy radiated per unit time per unit surface area at a temperature T.

Figure 18.3.3 shows what happens to the distribution curves as the temperature of a black body radiator is increased.

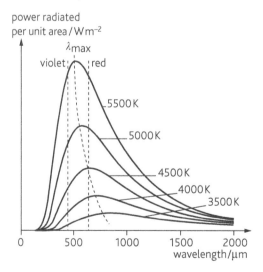

Figure 18.3.3 *Effect on spectrum as temperature increases*

From the graphs it can be seen that the distribution curve changes as the temperature of the black body changes. The proportion of the energy carried away by the shorter wavelengths increases with temperature. This can be said, because at higher temperatures, the distribution curve peaks at a lower wavelength. If a body is hot enough, it will begin emitting radiation in the visible region of the electromagnetic spectrum.

At around $1200\,\text{K}$, the visible wavelength that is emitted, lies mostly in the red end of the spectrum. At higher temperatures, the colour of the body changes from red to yellow to white.

There is a relationship between λ_{max} of each curve and its temperature.

It can be expressed as follows

$$\lambda_{max}T = \text{constant}$$

This relationship is known as Wien's displacement law.

Stefan's law

Definition

Stefan's law states that the rate at which energy is radiated from a black body is proportional to the fourth power of the temperature of the body expressed in kelvin.

The area under each curve gives the total energy E, radiated per unit time per unit surface area at a temperature T. **Stefan's law** states that the area under the graph is proportional the fourth power of the temperature of the body expressed in kelvin.

Stefan's law can be expressed mathematically as follows.

Equation

Stefan's Law

$P = \sigma A e T^4$

P	– energy radiated per second/W
σ	– Stefan's constant ($5.67 \times 10^{-8}\,\text{W}\,\text{m}^{-2}\,\text{K}^{-4}$)
A	– surface area/m^2
e	– emissivity of the body ($e = 1$ for a black body)
T	– surface temperature/K

When an object radiates energy, it also absorbs energy from the surroundings. The net energy emitted by an object is equal to the difference between the energy emitted by the object and the energy absorbed by the object. The net energy radiated per second is given by the equation below.

Equation

$P_{net} = \sigma A e (T^4 - T_s^{\,4})$

P_{net}	– net energy radiated per second/W
σ	– Stefan's constant ($5.67 \times 10^{-8}\,\text{W}\,\text{m}^{-2}\,\text{K}^{-4}$)
A	– surface area/m^2
e	– emissivity of the body ($e = 1$ for a black body)
T	– surface temperature/K
T_s	– temperature of surroundings/K

Example

A blackened sphere ($e = 0.25$) of radius $12\,\text{cm}$, initially at a temperature of $40\,°\text{C}$ is suspended inside a box which is held at a constant temperature of $110\,°\text{C}$.

a Calculate the rate of absorption of energy by the sphere inside the box.

b Calculate the initial net rate of absorption of energy by the sphere.

a $T = 273.15 + 110 = 383.15\,\text{K}$

Surface area of sphere $A = 4\pi r^2 = 4\pi(12 \times 10^{-2})^2 = 1.81 \times 10^{-1}\,\text{m}^2$

Rate of energy absorption

$P = \sigma A e T^4$

$= (5.67 \times 10^{-8})(1.81 \times 10^{-1})(0.25)(383.15)^4 = 55.25\,\text{W}$

b $T = 273.15 + 40 = 313.15\,\text{K}$

Net rate of absorption of energy

$P = \sigma A e (T^4 - T_s^4)$

$= (5.67 \times 10^{-8})(1.81 \times 10^{-1})(0.25)(383.15^4 - 313.15^4) = 30.5\,\text{W}$

Example

The emissivity of a person's skin is 0.74. Normal body temperature is 310 K. $0.37\,\text{m}^2$ of skin is exposed to the atmosphere which is at a temperature of 308K. Calculate the net energy radiated per second from the exposed skin.

$P_{\text{net}} = \sigma A e (T^4 - T_s^4)$

$= 5.67 \times 10^{-8} \times 0.37 \times 0.74 \times (310^4 - 308^4)$

$= 3.66\,\text{W}$

Key points

- Thermal radiation is the process by which thermal energy is transferred from a region of high temperature to a region of low temperature by electromagnetic waves.

- The rate at which thermal energy flows by radiation depends on temperature.

- When a body is in thermal equilibrium with another body, there is a flow of thermal energy back and forth between them.

- Good absorbers are also good emitters of thermal energy.

- A black body is a perfect absorber of thermal energy. The concept is a theoretical one.

- Stefan's law states that the rate at which energy is radiated from a black body is proportional to the fourth power of the temperature of the body expressed in kelvin.

Learning outcomes

On completion of this section, you should be able to:

- explain the greenhouse effect
- discuss applications of the transfer of thermal energy by conduction, convection and radiation.

The greenhouse effect

Energy from the Sun reaches the Earth in the form of light and short wavelength infrared radiation. The radiation passes easily through the atmosphere and heats up the Earth. This temperature of the Earth rises. This causes the Earth to start emitting infrared radiation. The wavelength of this infrared radiation is longer than that coming from the Sun. (The temperature of the Earth is much lower than that of the Sun.) Carbon dioxide in the atmosphere prevents the longer wavelength infrared radiation from leaving the Earth. The trapped energy causes the temperature of the Earth to rise even further. This process of heating up the Earth is known as the **greenhouse effect**. The burning of fossil fuels produces large volumes of carbon dioxide gas. This contributes significantly to **global warming**. The long term consequences of global warming are:

- melting of the polar ice caps resulting in increasing sea levels. The Caribbean is very prone to rising sea levels
- climate changes
- extinction of plant and animal life forms.

Applications of thermal energy transfer

The vacuum flask

The vacuum flask keeps hot liquids hot and cold liquids cold. The flask is designed such that it is difficult for thermal energy to travel into or out of the flask. The flask is a double-walled glass vessel with a vacuum between the walls. The vacuum eliminates heat loss by conduction and convection. The glass walls are silvered. The silvered surfaces reduce radiation entering from the outside of the flask. The cork stopper reduces the heat loss by conduction (Figure 18.4.1).

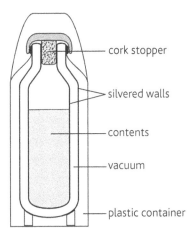

Figure 18.4.1 A vacuum flask

The solar water heater

Solar water heaters, as the name suggests, are used to heat water for use in homes and hotels. The solar water heater consists of an insulated box and an insulated storage tank. Inside the insulated box, there are copper pipes, which make thermal contact with a blackened surface. The insulated box is covered with either glass or plastic (Figure 18.4.2). Short wavelength infrared radiation enters through glass and heats up the insulated box. Longer wavelength infrared radiation cannot escape through the glass or plastic and gets trapped, warming up the inside of the box (greenhouse effect). The blackened surface is a good absorber of radiation. Solar radiation warms up the black surface and because the copper pipes (good conductor) are in contact with it, the water inside them heats up. The warm water is stored in an insulated storage tank for later use.

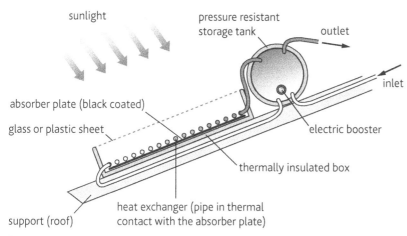

Figure 18.4.2 A solar water heater

Key points

- Short wavelength infrared radiation easily enters the atmosphere.
- The Earth absorbs some of this energy.
- The Earth then re-radiates longer wavelength infrared radiation, which cannot penetrate the atmosphere. Greenhouse gases such as carbon dioxide cause heat to be trapped.
- The temperature of the Earth rises and this is called global warming.
- The more carbon dioxide is pumped into the atmosphere, the more global warming takes place.

Answers to questions that require calculation can be found on the accompanying CD.

1 Copper and glass have different thermal conductivities. Explain this difference in terms of the mechanisms used to transfer thermal energy in each material. [4]

2 A copper rod of length 0.8 m and cross-sectional area 6.0×10^{-5} m² is arranged such that it is completely lagged on the sides. One end is maintained at 100 °C and the other end is maintained at 10 °C. Calculate the rate of flow of thermal energy through the copper rod. (Thermal conductivity of copper $= 390$ W m⁻¹K⁻¹) [2]

3 An ideally lagged compound bar consists of a copper bar 12 cm long joined to an aluminium bar of length 26 cm and of equal cross-sectional area. The free end of the copper bar is maintained at 120 °C and the free end of the aluminium is maintained at 0 °C. Calculate the temperature at the point at which the copper and aluminium bars are joined. [3]

(Thermal conductivity of copper $= 390$ W m⁻¹K⁻¹, thermal conductivity of aluminium $= 220$ W m⁻¹K⁻¹)

Sketch a graph to show the variation of temperature along the compound bar. [3]

4 Explain how land and sea breezes are formed. [6]

5 When determining the thermal conductivities of copper and glass, the choice of dimension used is very different. State what dimensions are used and why. [4]

6 Describe an experiment to find the thermal conductivity of copper in the form of a cylinder. [6]

7 Describe an experiment to find the thermal conductivity of glass. [6]

8 State the formula for linear heat flow by conduction explaining what the symbols mean. [3]

9 Explain what is meant by *thermal radiation*. [2]
Write an equation for Stefan's law of black body radiation and explain what each term means in the equation. [4]

10 A blackened sphere of radius 6 cm is in equilibrium with its surroundings. Calculate the temperature of the sphere if it absorbs 40 kW of power from the surroundings. [3]

11 a Explain what is meant by global warming. [3]
b What are the consequences of global warming? [3]
c Suggest three ways we can help reduce global warming. [3]

12 An electric iron is opened to expose the bare heating coils. The electric iron is switched on and a constant electrical power is supplied to it. The coils glow red hot.
a One of the ways by which thermal energy is lost from the heating coil is by convection. Name the other two methods by which thermal energy is lost. [2]
b Explain why the heating coil glows less brightly when some air is blown over it. [2]

13 Describe the principal features of a solar water heater and explain the design in terms of thermal energy transfer. [5]

Equation

Relative density
$$= \frac{\text{density of substance}}{\text{density of water}}$$

Density

The density of a substance gives an idea of how concentrated matter is. **Density** is defined as the mass per unit volume. The symbol used to represent density is ρ (rho). The SI unit of density is $kg\,m^{-3}$.

Definition	Equation
Density is defined as the mass per unit volume.	$\rho = \frac{m}{V}$ ρ – density/$kg\,m^{-3}$ m – mass/kg V – volume/m^3

The **relative density** of a substance is defined as the density of the substance divided by the density of water. Relative density has no units and is therefore dimensionless.

A kinetic model for solids, liquids and gases

A kinetic model may be used to explain the properties of solids, liquids and gases. Matter is thought to be made up of many tiny particles. These particles tend to attract each other. They also tend to move about. The term particle can refer to atoms, ions or molecules. The three common states of matter are solids, liquids and gases. Figures 19.1.1 (a), (b) and (c) show the typical arrangement of the particles that make up a solid, liquid and a gas.

Solids

In a solid, the particles are held closely together by strong forces of attraction. The spacing between the particles is therefore very small. The particles are arranged in an orderly manner. The particles are not free to move about within the solid. Instead, they vibrate about their fixed positions. As a result of this, solids have fixed shapes and volumes.

Liquids

In a liquid, the force of attraction between the particles is weaker than that of a solid. The separation between the particles is slightly more than in a solid. There is less ordering in a liquid than in a solid. The particles

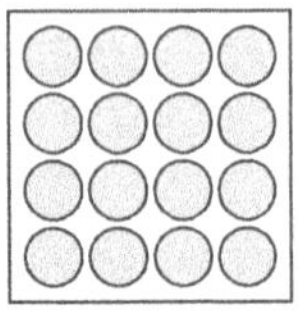

spacing – very small
ordering – well ordered arrangement of particles
motion of particles – free to vibrate about their fixed positions

Figure 19.1.1 (a) Structure of a solid

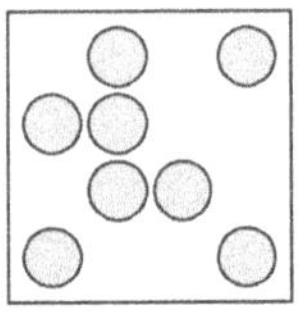

spacing – very small
ordering – well ordered arrangement of particles
motion of particles – free to vibrate about their fixed positions

Figure 19.1.1 (b) Structure of a liquid

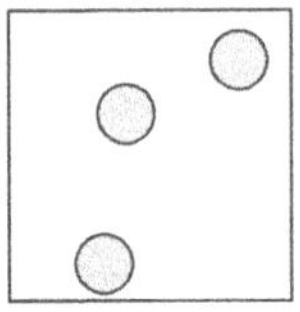

spacing – very large
ordering – particles randomly distributed; there is no order
motion of particles – moving randomly at high speeds

Figure 19.1.1 (c) Structure of a gas

are able to move throughout the body of the liquid. As a result, liquids take the shape of the container they are in and their volumes are fixed.

Gases

In a gas, the particles are far apart from each other when compared with a solid or liquid. The particles move about randomly at high speeds, colliding with each other and the walls of the container holding the gas. Since the particles are far apart from each other, the force of attraction between them is negligible. As a result, a gas fills up any space that is available.

Brownian motion

The apparatus used to demonstrate Brownian motion is shown in Figure 19.1.2. It consists of a small transparent cell containing some smoke. The cell is illuminated from the side and the smoke particles are observed using a microscope. The microscope is adjusted so that the smoke particles are seen as bright specks. The bright specks are seen moving around haphazardly. The haphazard motion of the smoke particles is called **Brownian motion**.

Explanation – The movement of the smoke particles is due to air molecules bombarding them from all sides. The haphazard motion of the smoke particles suggests that the air molecules are moving around rapidly in all directions.

If larger smoke particles are used, the bright specks are seen moving less haphazardly because the randomness of the collisions averages out.

Brownian motion provides evidence for the kinetic model of matter.

Structure of crystalline and non-crystalline solids

In a **crystalline solid**, the atoms, ions or molecules are arranged in a regular pattern that repeats itself within the crystal. An example of a crystalline solid is quartz. Metals are also crystalline solids. Figure 19.1.3 shows a simple structure of a crystalline solid.

In an **amorphous** solid, there is a random arrangement of the atoms throughout the structure. There is no definite pattern. An example of an amorphous material is glass. Glass is a mixture of silica and other substances. The atoms in this case are arranged to form an amorphous structure.

Polymers consist of long chains of molecules. Each chain consists of units that repeat themselves. Each repeating unit is called a **monomer** (Figure 19.1.4).

Polymers can be semi-crystalline or amorphous. The long polymer chain may be coiled up or tangled. The amount of tangling determines whether the polymer is semi-crystalline or amorphous. Figure 19.1.5 and Figure 19.1.6 below show the difference between the two structures.

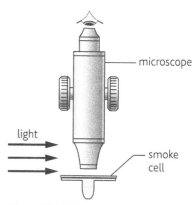

Figure 19.1.2 *Brownian motion*

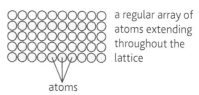

Figure 19.1.3 *The structure of a crystalline solid*

Figure 19.1.4 *A polymer chain*

Key points

- Density is defined as the mass per unit volume.
- Relative density is the ratio of the density of a substance to the density of water. It has no units.
- A kinetic model can be used to explain the properties of solids, liquids and gases.
- Brownian motion demonstrates that molecules are in continuous random motion.
- In a crystalline solid, the atoms, ions or molecules are arranged in a regular pattern that repeats itself within the crystal
- In an amorphous solid, there is a random arrangement of the atoms throughout the structure.
- Polymers consist of long chains of molecules. Each chain consists of units that repeat themselves. Each repeating unit is called a monomer.
- Polymers can be semi-crystalline or amorphous.

Figure 19.1.5 *A semi-crystalline polymer*

Figure 19.1.6 *An amorphous polymer*

19.2 Pressure

Learning outcomes

On completion of this section, you should be able to:

- define pressure
- derive an expression for the pressure in a fluid
- recall the various instruments used to measure pressure.

Definition

Pressure is defined as the force acting normally per unit area.

Equation

$$p = \frac{F}{A}$$

p – pressure/Pa
F – force/N
A – area/m²

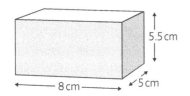

Figure 19.2.1

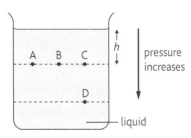

Figure 19.2.2

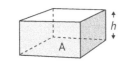

Figure 19.2.3

Pressure

The **pressure** acting on a surface is defined as the force exerted normally per unit area on the surface. The SI unit of pressure is the pascal (Pa). From the definition it follows that $1\,\text{Pa} = 1\,\text{N m}^{-2}$.

A block of wood of mass 2 kg rests on a table as shown in Figure 19.2.1. Calculate the pressure exerted on the surface of the table.
[Use g = 9.81 N kg⁻¹]

$$\text{Area in contact with the surface of the table} = l \times b$$
$$= 5 \times 10^{-2} \times 8 \times 10^{-2}$$
$$= 0.004\,\text{m}^2$$

$$\text{Force exerted on table} = mg$$
$$= 2 \times 9.81 = 19.62\,\text{N}$$

$$\therefore \quad \text{pressure} = \frac{\text{force}}{\text{area}} = \frac{19.62}{0.004} = 4.91 \times 10^3\,\text{Pa}$$

Pressure in a fluid

A fluid is a substance that has the ability to flow. Liquids and gases are fluids. Above the Earth's surface, there is a thick layer of gas. This layer is called the atmosphere. The atmosphere exerts a pressure of 1.01×10^5 Pa on the Earth's surface. Even though this pressure seems large, it is usually not felt by us. This is because the insides of our body exert an outward pressure. This prevents our bodies from collapsing under the pressure exerted by the atmosphere. The pressure becomes more significant on a diver at the bottom of a lake. In this case the diver is experiencing atmospheric pressure as well as the pressure exerted by the water in the lake above him.

In a fluid, pressure increases with depth. At a particular depth, pressure acts equally in all directions.

In Figure 19.2.2, a container is filled with a liquid. The pressure in the liquid is greatest at the bottom of the container. The points A, B and C are at the same depth in the liquid and are therefore at the same pressure. The point D is deeper than A, B and C, and is therefore at a higher pressure.

Consider a container of height h, filled with a fluid of density ρ. The cross-sectional area of the base of the container is A (Figure 19.2.3).

$$\text{Volume of fluid in the container} = \text{cross-sectional area} \times \text{height} = Ah$$

$$\text{Mass of fluid} = \text{density} \times \text{volume} = \rho Ah$$

$$\text{Weight of fluid} = \text{mass of fluid} \times \text{gravitational field strength} = \rho Ahg$$

$$\text{Therefore, the force exerted on A} = \rho Ahg$$

$$\text{Pressure exerted on the surface A} = \frac{\text{force}}{\text{area}} = \frac{\rho Ahg}{A} = \rho gh$$

It follows that the difference in pressure Δp, between two points separated by a vertical distance Δh in a fluid of density ρ is given by $\rho g \Delta h$. The expression indicates that the pressure at any depth is independent of the area of the base. This simply means that if a cylinder was used in the derivation the same expression would have been obtained.

Example

A diver is 10 m below the surface of a lake. Calculate the pressure exerted on the diver at this depth.

[Density of water = $1 \times 10^3\,kg\,m^{-3}$, atmospheric pressure = $1 \times 10^5\,Pa$, $g = 9.81\,N\,kg^{-1}$]

Pressure due to water $= \rho g h$

$$= 1 \times 10^3 \times 9.81 \times 10 = 9.81 \times 10^4\,Pa$$

Total pressure acting on diver

$$= \text{pressure due to atmosphere} + \text{pressure due to water}$$

$$= 1 \times 10^5 \times 9.81 \times 10^5 = 1.981 \times 10^5\,Pa$$

Instruments used to measure pressure

The U-tube manometer

This device consists of a U-shaped tube containing a liquid of known density ρ. Mercury is commonly used. The U-tube manometer can be used to measure the pressure of a gas. One end of the tube is connected to the container holding the gas. The other end is open to the atmosphere. The pressure at the point A is atmospheric pressure P_A. The pressure at the point B is the pressure exerted by the liquid column of height h plus atmospheric pressure (Figure 19.2.4). The pressure of the gas is actually the same as the pressure at the point B.

The pressure of the gas is given by $p = P_A + \rho g h$.

The Bourdon gauge

The Bourdon gauge consists of a curved metal tube. The tube is closed at one end, which is connected to a pointer. When a gas is connected to the gauge, the pressure causes the tube to straighten out slightly causing the pointer to move. The scale is calibrated to measure pressure directly (Figure 19.2.5).

The barometer

A barometer consists of a long glass tube partially filled with mercury. It is inverted in a tray, also containing mercury. The barometer is used to measure atmospheric pressure. Depending on the atmospheric pressure, the height of the mercury in the glass tube will either increase or decrease (Figure 19.2.6). The atmospheric pressure is given by

$$P_A = \rho g h$$

where

ρ is the density of mercury, $\quad g$ is the gravitational field strength,

h is the height of mercury in the column

Key points

- Pressure is defined as the force acting normally per unit area.
- The pressure in a fluid depends on depth, gravitational field strength and density.
- Pressure can be measured using U-tube manometer, barometer or a Bourdon gauge.

Equation

$$\Delta p = \rho g \Delta h$$

Δp – difference in pressure/Pa
ρ – density of fluid/kg m^{-3}
g – gravitational field strength/N kg^{-1}
Δh – difference in height/m

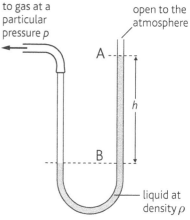

Figure 19.2.4 A U-tube manometer

Figure 19.2.5 A Bourdon gauge

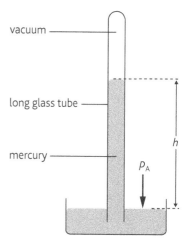

Figure 19.2.6 A barometer

19.3 Hooke's law and the Young modulus

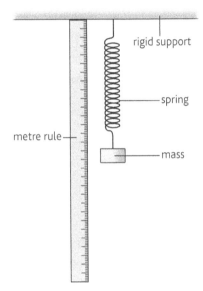

Figure 19.3.1 *Demonstrating Hooke's law*

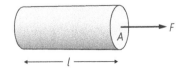
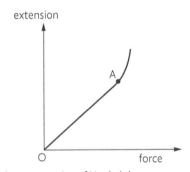

Figure 19.3.3 *Defining tensile stress*

Hooke's law

A spring is made by making a coil from a piece of metal. In order to change the shape of the spring, a force is required. If the spring is lengthened by stretching, the applied force is called a tensile force. If the spring is shortened by pressing, the applied force is called a compressive force. When a tensile force is applied to a spring its length increases. The increase in length is called an extension. A spring is attached to a rigid support (Figure 19.3.1). A mass is attached to it and the extension produced is measured using a metre rule. Suppose additional masses are attached to the spring and the extension produced in each case is measured.

The data obtained can be used to plot a graph of extension against force. Force is the manipulated variable and is therefore plotted on the x-axis. Extension is the variable being measured as a result of a change in force and is therefore plotted on the y-axis. Figure 19.3.2 shows a typical graph that is obtained by stretching a spring.

extension

O   force

Figure 19.3.2 *Graphical representation of Hooke's law*

From the graph you can seen that there is a straight line portion between O and A. In this region, the extension is proportional to the force. This can be expressed mathematically as

$$F = ke$$

The term k, in the expression is called the spring constant. It is defined as the force per unit extension. Its unit is $\mathrm{N\,m^{-1}}$.

In the region OA, the spring is said to obey Hooke's law.

Definition

Hooke's law states that the extension produced is proportional to the force applied, provided that the elastic limit is not exceeded.

Definition

The spring constant k is defined as the force acting per unit extension.

Stress, strain and the Young modulus

Consider a cylindrical piece of material of length l, having a cross-sectional area of A. Suppose a force F is applied on one end of the material (Figure 19.3.3). This force is called a tensile force because it is stretching the material. The **stress** exerted on the material is the force acting normally per unit cross-sectional area. The SI unit for stress is the pascal (Pa) or $\mathrm{N\,m^{-2}}$.

Definition	*Equation*
Stress is defined as the force acting per unit cross-sectional area.	$\text{stress} = \dfrac{F}{A}$ F – force/N A – cross-sectional area/m² Unit of stress /Pa or Nm⁻²

Suppose the force F causes the material to produce an extension e (Figure 19.3.4). The **strain** exerted on the material is the ratio of the extension produced to the original length of the material. Strain has no units because it is a ratio of lengths.

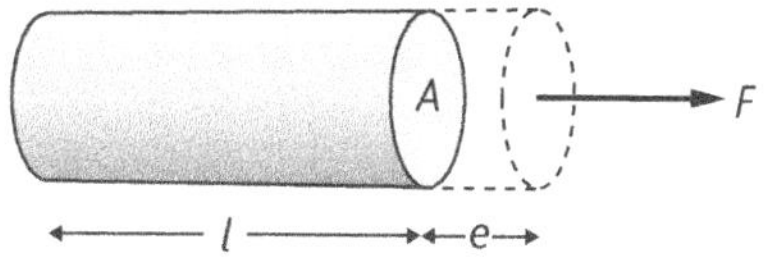

Figure 19.3.4 *Defining tensile strain*

Definition	*Equation*
Strain is defined as the ratio of the extension to the original length of the material.	$\text{strain} = \dfrac{e}{l}$ e – extension/m l – original length/m Strain is a ratio of lengths and has no units

Another quantity of importance is the **Young modulus**. This is the ratio of the tensile stress to the tensile strain. The SI unit for the Young modulus is also the pascal (Pa) or Nm⁻².

Equation

$\text{Young modulus } (E) = \dfrac{\text{tensile stress}}{\text{tensile strain}}$

E – Young modulus /Pa or Nm⁻²

Example

A metal wire of length 2.0 m is clamped vertically. A load is attached to and it extends 0.56 mm. The cross-sectional area of the wire is $1.1 \times 10^{-7}\,\text{m}^2$.

Calculate

i   the strain in the wire
ii the force applied to the wire.

[Young modulus of wire $= 2.0 \times 10^{11}\,\text{Pa}$]

i $\text{Strain} = \dfrac{e}{l} = \dfrac{0.56 \times 10^{-3}}{2.0} = 2.8 \times 10^{-4}$

ii $E = \dfrac{F/A}{e/l} = \dfrac{Fl}{Ae}$

$F = \dfrac{EAe}{l} = EA \times \text{strain} = (2 \times 10^{11} \times 1.1 \times 10^{-7}) \times 2.8 \times 10^{-4}$

$\qquad\qquad = 6.16\,\text{N}$

Key points

- Hooke's law states that the extension produced is proportional to the force applied, provided that the elastic limit is not exceeded.
- The spring constant k is defined as the force acting per unit extension.
- Stress is defined as the force acting per unit cross-sectional area.
- Strain is defined as the ratio of the extension to the original length of the material.
- The Young modulus is the ratio of the tensile stress to the tensile strain.

On completion of this section, you should be able to:

- describe an experiment to determine the Young modulus of a metal in the form of a wire.

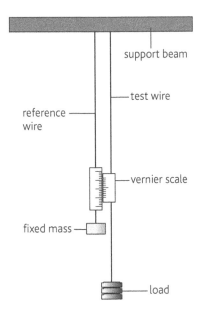

reference wire

support beam

test wire

vernier scale

fixed mass

load

Figure 19.4.1 *Experiment to determine the Young modulus of a metal*

Experiment to measure E

The Young modulus of a material like copper can be determined by using a length of copper wire. The extension produced when a load is applied to a piece of wire can be very small. In order to produce measurable extensions in this experiment, a piece of wire having a length of at least one metre is used.

In this experiment (Figure 19.4.1), two pieces of wire are attached to a rigid support, adjacent to each other. One wire is used as a reference wire and the other is used as the test wire. The reference wire has a mass attached to the end to keep the wire vertical. The reference wire is needed, because it compensates for temperature changes and sagging of the supporting beam.

The diameter of the test wire, d, is measured using a micrometre screw gauge. Several measurements are taken at different positions on the wire and the average value is determined. This reduces the random error in the value of d.

The original length of the test wire l, is measured using a metre rule or a measuring tape.

A load is first measured using a balance. It is then attached to the test wire and the extension is measured using a Vernier scale.

The experiment is repeated several times, using increasing loads. The experiment is also repeated when unloading the masses from the test wire. This can be used to determine if the elastic limit was exceeded.

Measurements to be taken:

- The original length of the wire l is measured using a metre rule or tape.
- The diameter of the wire d is measured using a micrometre screw gauge.
- The mass of the each load is measured using a balance.
- The extension of the wire is measured using a vernier scale.

Determination of the Young modulus

The results of the experiment can be represented as follows:

Original length of wire $\quad= l\,\mathrm{m}$

Diameter of wire $\quad\quad\quad= d\,\mathrm{m}$

Cross-sectional area of wire $\quad= \pi\left(\dfrac{d}{2}\right)^2 = \dfrac{\pi d^2}{4}\,\mathrm{m}^2$

The results can be entered in a table such as the following.

Table 19.4.1

	m/kg	F = mg/N	Extension/m (loading)	Extension/m (unloading)
1				
2				
3				
4				
5				
6				

A graph of force against extension is plotted (Figure 19.4.2) using the results in the table.

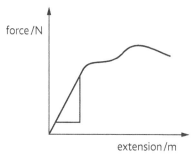

Figure 19.4.2 Plot a graph of force against extension

The gradient of the *linear region* of the graph is determined.

$$\text{Young modulus} = \frac{\text{stress}}{\text{strain}} = \frac{F}{A} \div \frac{e}{l} = \frac{Fl}{eA}$$

The gradient of the graph is $\frac{F}{e}$.

Therefore, the Young modulus $=$ gradient $\times \frac{l}{A}$

Example

Figure 19.4.3 shows the results of an experiment using a metal wire with length 2.0 m and diameter 1.1 mm.

$$\text{Cross-sectional area of wire} = \pi\left(\frac{d}{2}\right)^2 = \pi\left(\frac{1.1 \times 10^{-3}}{2}\right)^2$$
$$= 9.5 \times 10^{-7} \, \text{m}^2$$

$$\text{Gradient of straight line} = \frac{42}{0.6 \times 10^{-3}} = 7.0 \times 10^4 \, \text{N m}^{-1}$$

$$\text{Young modulus} = \text{gradient of straight line} \times \frac{l}{A}$$

$$= 7.0 \times 10^4 \times \frac{2.0}{9.5 \times 10^{-7}} = 1.47 \times 10^{11} \, \text{Pa}$$

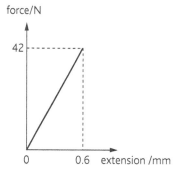

Figure 19.4.3

Definition

Elastic deformation – the material returns to its original shape and size when the external force is removed.

Definition

Plastic deformation – the material does not return to its original shape and size when the external force is removed from it.

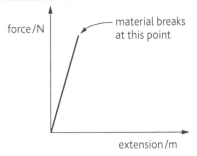

Figure 19.5.2 *Force–extension graphs for brittle materials – (glass and cast iron)*

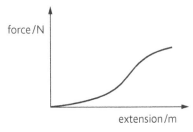

Figure 19.5.3 *Force–extension graphs for polymeric materials – (rubber and elastic)*

Force–extension graphs

Ductile materials

A **ductile** material can be easily stretched and formed into a wire. An example of a ductile material is copper. Figure 19.5.1 shows a typical force-extension graph for a ductile material. In the region OA, Hooke's law is obeyed. The point A is called the limit of proportionality. Beyond this point Hooke's law is not obeyed and the graph is no longer linear. The point B is called the elastic limit. This is the maximum load which a body can experience and still regain its original size and shape, once the load is removed. When the force in increased beyond the elastic limit, a point is reached where there is a marked increase in extension. The point C is called the yield point. The point D is the point at which the material breaks. The graph also illustrates the regions where there is **elastic deformation** (O to B) and **plastic deformation** (B to D). In the elastic region, the material returns to its original shape and size when the force is removed. In the plastic region, the material does not return to its original shape and size when the force is removed.

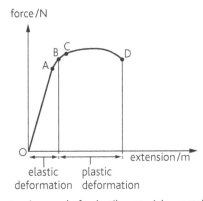

Figure 19.5.1 *Force–extension graphs for ductile materials – metal (copper)*

Brittle materials

Glass and cast iron are good examples of brittle materials. The force–extension graph for brittle materials is a straight line with a steep gradient (Figure 19.5.2). There is no plastic region on the graph. There is only an elastic region. This means that once the force is removed, the material will return to its original shape and size.

Polymeric materials

An example of a polymeric material is rubber. The force–strain graph is usually non-linear (Figure 19.5.3). This means that it does not obey Hooke's law. A small force produces a large extension.

Stress–strain graphs

Strain–strain graphs are similar to force–extension graphs. Figures 19.5.4–6 show the effects applying and removing loads on ductile, brittle and polymeric materials. In the case of the ductile material, the material does not return to its original length when the load is removed. The material has been permanently deformed.

In the case of the brittle material, the material will return to its original shape and size if the ultimate tensile stress is not exceeded.

In the case of the polymeric material, the material returns to its original shape and size. It should be pointed out that during the loading and unloading process, energy is lost as heat. The effect is known as elastic hysteresis. The area between the two curves represents the energy lost per unit volume. The rubber actually gets warmer in the process.

Definition

The ultimate tensile stress is the maximum stress a material can withstand. It is shown as the maximum point on a stress–strain curve.

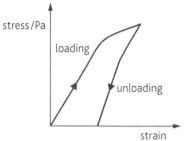

Figure 19.5.4 A ductile material

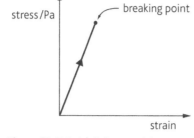

Figure 19.5.5 A brittle material

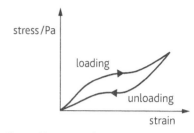

Figure 19.5.6 A polymeric material

Strain energy

Figure 19.5.7 shows how the extension varies with force for a material that obeys Hooke's law. The work done for an extension of e is given by the shaded area. The area of the triangle is $\frac{1}{2} \times$ base $\times$ height.

$$\therefore \qquad \text{work done} = \frac{1}{2}Fe$$

Since work is being done on the material, energy is stored in the material as strain energy.

$$\therefore \qquad \text{strain energy} = \frac{1}{2}Fe$$

Example

A thin steel wire initially of length 1.2 m and diameter 0.6 mm is suspended from the ceiling. A mass of 4.0 kg is attached to the lower end of the wire. Calculate:

i the extension produced in the steel wire

ii the energy stored in the wire.

State an assumption you made in your calculation.

[Young modulus for steel $= 2.0 \times 10^{11}$ Pa, $g = 9.81$ N kg^{-1}]

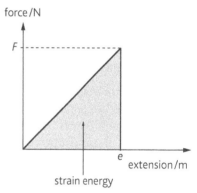

Figure 19.5.7 Strain energy

i $E = \dfrac{F/A}{e/l} = \dfrac{Fl}{Ae}$

Force exerted on wire $= 4.0 \times 9.81$ N

Cross-sectional area of wire $= \pi \left(\dfrac{d}{2}\right)^2 = \pi \left(\dfrac{0.6 \times 10^{-3}}{2}\right)^2$ m^2

$E = \dfrac{F/A}{e/l} = \dfrac{(4.0 \times 9.81)(1.2)}{\pi \left(\dfrac{0.6 \times 10^{-3}}{2}\right)^2 (2.0 \times 10^{11})} = 8.3 \times 10^{-4}$ m

ii Energy stored in the wire $= \dfrac{1}{2}Fe = \dfrac{1}{2} \times (4.0 \times 9.81) \times (8.3 \times 10^{4})$

$= 1.63 \times 10^{-2}$ J

In the calculations, it is assumed that Hooke's law is obeyed.

Key points

- Elastic deformation – the material returns to its original shape and size when the external force is removed from it.

- Plastic deformation – the material does not return to its original shape and size when the external force is removed from it.

- When a material is stretched, external work is done on the material. The energy stored in the material is stored as strain energy.

Revision questions 10

Answers to questions that require calculation can be found on the accompanying CD.

1 a Distinguish between the structure of crystalline and non-crystalline solids. [2]
  Give an example of each. [2]

  b Describe the structure of a polymer. Give one example of a commonly used polymer. [3]

2 a Explain what is meant by Brownian motion. [2]

  b Describe how you would demonstrate Brownian motion. [3]

  c Interpret the results of such an experiment. [2]

3 Distinguish between a solid and a liquid by reference to the spacing, ordering and motion of the particles. [3]

4 a Define the term *pressure* and state its SI unit. [2]

  b Derive an expression for the pressure exerted by a fluid. [4]

5 a Define the Young modulus of a material. [2]

  b A thin steel wire initially of length 1.4 m and diameter 0.5 mm is suspended from the ceiling. A mass of 3.5 kg is attached to the lower end of the wire.
  Calculate
  i the extension produced in the steel wire [3]
  ii the energy stored in the wire. [2]
  State an assumption you made in your calculation.
  [Young modulus for steel $= 2.0 \times 10^{11}$ Pa, $g = 9.81$ N kg^{-1}]

6 a State Hooke's law. [2]

  b Explain what is mean by the term *spring constant*. [1]

  c A spring is compressed by applying a force to it. The variation of this force F with compression x is shown below.

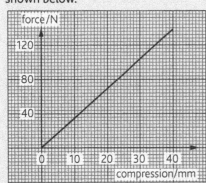

Calculate:
  i the spring constant [2]
  ii the work done in compressing the spring by 35 mm. [2]

7 a Explain what is meant by the terms:
  i stress [1]
  ii strain [1]
  iii Young modulus. [1]

8 Some bridges are designed such that many cables, made of high-tensile steel, are used to support the roadway. A particular cable has a cross-sectional area of 8.0×10^{-3} m^2 and length 40 m. The tension in this cable is 650 kN.
[Young modulus of the steel $= 2.8 \times 10^{11}$ Pa]
Calculate:
  i the extension of the cable [5]
  ii the strain energy stored in the cable. [2]

9 a Describe, with the aid of a diagram, the apparatus that can used to measure the Young modulus of a length of copper wire. [4]

  b Describe the method used to determine the required measurements. [4]

  c Describe how the measurements taken can be used to determine the Young modulus. [4]

10 a Define the terms *stress* and *strain*. [2]

  b Distinguish between elastic and plastic deformation. [2]

11 Give an example and sketch a force–extension graph for each of the following:
  a a brittle material [2]
  b a polymeric material [2]
  c a ductile material. [2]
  For each material, use the terms *plastic* and *elastic* to describe their behaviour as indicated by the graphs. [6]

12 Give an example of each of these materials.
  i a ductile material [1]
  ii a brittle material [1]
  iii a polymeric material. [1]
  For each of these materials, sketch a possible stress–strain graph. [3]

13 Define the following terms:

    a limit of proportionality   [1]

    b elastic limit   [1]

    c yield point.   [1]

14 Aluminium has a Young modulus of 7.1×10^{10} Pa. A piece of aluminium wire of length 1.5 m is attached to a rigid support. The diameter of the wire is 1.5 mm. A load of 30 N is attached to the free end.

Calculate:

    a the stress on the wire   [2]

    b the extension produced   [2]

    c the strain energy stored in the wire.   [2]

15 A light rigid bar is suspended horizontally by a brass and steel wire as shown below.

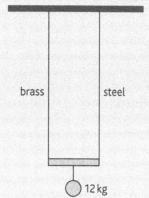

The length of each wire is 2.2 m. When a mass of 12 kg is attached to the bar, it remains horizontal.

[Young modulus of brass $= 1.0 \times 10^{11}$ Pa

Young modulus of steel $= 2.0 \times 10^{11}$ Pa]

    a Calculate the tension in each wire.   [2]

    b Given that the cross-sectional area of the brass wire is 6.0×10^{-7} m^2, calculate the extension of the brass wire.   [3]

    c Calculate the energy stored in the brass wire   [2]

    d Determine the diameter of the steel wire.   [3]

Module 3 Practice exam questions

Answers to the multiple-choice questions and to selected structured questions can be found on the accompanying CD.

Multiple-choice questions

1. Which of the following thermometers can be used to measure the temperature at various positions of a flame produced by a Bunsen burner?
   a A platinum-resistance thermometer
   b A thermocouple
   c A mercury-in-glass thermometer
   d A constant-volume gas thermometer

2. Which of the following is not an assumption of the kinetic theory of gases?
   a The size of the molecules is negligible when compared the volume occupied by the gas.
   b The force of attraction between the molecules is negligible.
   c The collisions are inelastic.
   d The duration of a collision is negligible when compared to the time between collisions.

3. A car has a kinetic energy of 55 kJ. The car is brought to rest by applying the brake. 60% the kinetic energy is converted into thermal energy in the brakes of the car. The total mass of the brake components is 25 kg and their average specific heat capacity is 510 J kg^{-1} K^{-1}. What is the increase in temperature of the brake components?
   a 4.3 °C   b 0.004 °C   c 2.6 °C   d 0.23 °C

4. An insulated composite rod of length 80 cm consists of two metals P and Q. P and Q are of the same lengths. The thermal conductivities of P and Q are 400 W m^{-1} K^{-1} and 300 W m^{-1} K^{-1} respectively. One end of P is at a temperature of 90 °C and one end of Q is at 12 °C. What is the temperature at the point where P and Q are joined?
   a 62 °C   b 57 °C   c 48 °C   d 51 °C

5. According to Stefan's law, the rate at which energy is transferred from a black body is given by which equation?
   a $P = \sigma AT^2$ b $P = \sigma AT$
   c $P = \sigma AT^4$ d $P = \sigma TA^4$

6. Which of the following statements is true about the evaporation and boiling of water?
   a Boiling occurs at any temperature and evaporation occurs 100 °C
   b Evaporation occurs at any temperature and boiling is dependent on the external pressure.
   c The rate of evaporation and boiling are unaffected by changes in the surface area of the water.
   d Evaporation and boiling occur only at the surface of the water.

7. What is the pressure of a gas of density 0.12 kg m^3 and root mean square velocity of 1200 m s^{-1}?
   a 48 Pa   b 5.76 × 10^4 Pa
   c 144 Pa   d 2.5 × 10^5 Pa

8. The first law of thermodynamics can be expressed as $\Delta U = \Delta Q + \Delta W$. When a gas undergoes an isothermal change, there is no change in temperature of the gas. The first law of thermodynamics for an isothermal change is
   a $\Delta U = \Delta W$ b $\Delta U = \Delta Q$
   c $-\Delta Q = \Delta W$ d $\Delta Q = \Delta W - \Delta U$

9. A force–extension graph of wire is shown below. What is the amount of work done in stretching the wire from extension 2 mm to 3 mm?
   a 0.004 J
   b 0.001 J
   c 0.009 J
   d 0.005 J

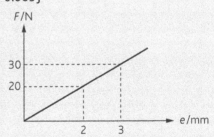

10. The length l of a material X is measured when various forces F are applied to it. The graph below shows the results of this investigation. Which of the following is false?
    a The material obeys Hooke's law up to 30 N.
    b The force constant of the material is 750 N m^{-1}.
    c The ultimate tensile stress is at 35 N.
    d The work done in stretching the material from 1.22 m to 1.24 m is 0.65 J

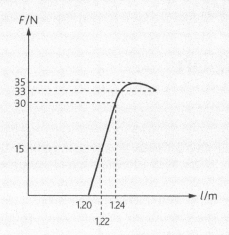

Structured questions

11 a Explain what is meant by temperature. [1]

b Explain how a physical property of a substance which varies with temperature may be used to measure the temperature of a substance. [2]

c Describe the principal features of a platinum-resistance thermometer. [4]

d Discuss the advantages and disadvantages of a platinum-resistance thermometer and a mercury-in-glass thermometer which may be used in the same temperature range. [4]

e A student measures the temperature of a liquid using a platinum-resistance thermometer and a mercury-in-glass thermometer. The student notices that the temperatures do not agree. They do, however, agree when used to measure the temperature of steam above pure boiling water. Explain why this is so. [3]

12 a Explain what is meant by the following:

   i Conduction   **ii** Convection
   iii Radiation. [6]

b Explain what is meant by thermal conductivity and state its SI unit. [3]

c Describe and state the mechanism for thermal energy transfer in the following:

   i a metal spoon   **ii** a ceramic cup. [6]

d State and explain the choice of dimensions the material used to experimentally determine the thermal conductivity of:

   i a good conductor   **ii** a poor conductor. [4]

13 a Explain what is meant by the following:

   i specific heat capacity of copper [2]
   ii specific latent heat of fusion of ice. [2]

b Describe an experimental method to measure the specific heat capacity of a liquid. [6]

c A kettle is rated at 2.0 kW. A mass of 800 g of water at 25 °C is poured into a kettle. The kettle is switched on and it takes 2.3 minutes for the water to start boiling. The kettle is left on and it takes 8.6 minutes for the mass of water to decrease by half its original value. Use the data to calculate:

   i the specific heat capacity of water [3]
   ii the specific latent heat of vaporisation of water. [2]
   iii State one assumption used in your calculations. [1]

14 a State four assumptions of the kinetic theory. [4]

b Explain the following using the kinetic theory:

   i the air inside a tyre exerts a pressure on the inner walls [4]
   ii the air pressure inside the tyre increases after a car has been driven [3]
   iii the air pressure decreases when some air escapes from the tyre. [2]

b Three molecules have speeds v, $2v$ and $4v$. Calculate:

   i the mean square speed of the molecules [2]
   ii the root mean square speed of the molecules. [2]

c 2.4 mol of a gas is heated at a constant pressure of 4.0×10^5 Pa. The temperature of the gas increases from 250 K to 300 K.

Molar heat capacity of the gas at constant pressure is 29 J mol^{-1} K^{-1}. Calculate

   i the amount of energy supplied to the gas [2]
   ii the change in volume of the gas [2]
   iii the work done by the gas. [2]

15 a Explain what is meant by:

   i tensile stress [1]
   ii tensile strain [1]
   iii Young modulus. [1]

b Sketch stress–strain graphs for the following and give an example of each type.

   i A brittle material [2]
   ii A ductile material [2]
   iii A polymeric material [2]

20.1 Analysis and interpretation

Learning outcomes

On completion of this section, you should be able to:

- draw straight line graphs using experimental data

- determine gradients and intercepts

- rearrange equations to obtain a straight line form.

Plotting graphs

In experimental work, data is often collected. This data must be analysed. One way of analysing data is by drawing a suitable graph to determine unknown constants.

The following guidelines should be used when drawing graphs.

- Choose a scale that is suitable (e.g. 1 cm to 5 units, 1 cm to 10 units). Do not use scales such as 1 cm to 3 units.

- The scale should be chosen such that the points being plotted occupy at least half the graph paper.

- The x- and y-axes should be labelled with the quantities being plotted and the units (e.g. d/cm, T/s, height/m, temperature/°C).

- The graph should be given an appropriate title.

- A × or ● a should be used to represent the points being plotted.

- If the data appear to follow a linear relationship, a line of best fit should be drawn so that there are equal numbers of data points on either side of the line.

- If the data appear to follow a non-linear relationship, a smooth curve should be drawn through the data points. Do not use a straight line between adjacent points.

Two quantities are required when plotting a graph (x-axis and y-axis). In an experiment to investigate the relationship between two quantities, only one quantity can be changed at a time. This quantity is called the manipulated variable. It is usually plotting on the x-axis. The second quantity is called the responding variable. The responding variable is usually plotted on the y-axis. Since only one quantity is being changed in the experiment, all other quantities are known as constant variables.

The general equation of a straight line is $y = mx + c$. The gradient or slope of the line is m. The intercept on the y-axis is c. Consider Figure 20.1.1.

A and B lie on the straight line. The coordinates of A and B are (x_1, y_1) and (x_2, y_2) respectively. The gradient of the line passing through A and B is given by:

$$\text{Gradient (slope)} = \frac{y_2 - y_1}{x_2 - x_1}$$

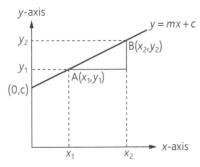

Figure 20.1.1 A straight-line graph

Example

A system is oscillating with simple harmonic motion. The period of oscillation T is dependent on the length of the oscillator x. It is known that T and x are related by the following equation:

$$T = Ax^{-b}$$

The following table shows the results of an experiment where the length of the oscillator is varied and the period of oscillation is measured in each case. It is required that the constants A and b be determined.

x/m	0.40	0.50	0.60	0.70	0.80	0.90
T/s	2.85	2.32	2.15	2.01	2.01	1.90

The equation has to be rearranged such that a linear form is obtained.

$$T = Ax^{-b}$$

Taking $\log_{10}$ on both sides of the equation:

$$\lg T = \lg (Ax^{-b})$$
$$\lg T = \lg A + \lg (x^{-b})$$
$$\lg T = \lg A - b \lg x$$

A straight line graph would be obtained if $\lg T$ is plotted against $\lg x$.

The gradient of the line is $-b$.

The y-intercept is $\lg A$.

$\lg T$ and $\lg x$ are determined as shown in the following table.

x/m	0.40	0.50	0.60	0.70	0.80	0.90
T/s	2.85	2.32	2.15	2.01	2.01	1.90
$\lg (x/m)$	−0.397	−0.301	−0.222	−0.155	−0.097	−0.046
$\lg (T/s)$	0.455	0.407	0.365	0.332	0.303	0.279

A line of best fit is drawn.

The graph of $\lg T$ against $\lg x$ is shown in Figure 20.1.2.

In order to find the gradient of the line, two points A and B are chosen on the line. These points are not one of the experimental data points and should be sufficiently far apart to produce a large right-angled triangle.

A – (−0.38, 0.45)

B – (−0.065, 0.285)

$$\text{Gradient} = -b = \frac{0.285 - 0.45}{-0.065 - (-0.38)} = -0.52$$

Therefore, $b = 0.52$

From the graph, the y-intercept $= 0.25$

$$\lg A = 0.25$$

Therefore, $A = 10^{0.25} = 1.78$

In the event that the y-intercept cannot be determined by direct measurement from the graph, a point on the line is chosen. Suppose the point A is chosen. The y-intercept is calculated as follows:

$$\lg T = \lg A - b \lg x$$
$$0.45 = \lg A - 0.52 (-0.38)$$
$$\lg A = 0.252$$
$$A = 10^{0.252} = 1.79$$

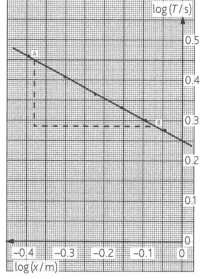

Figure 20.1.2

Answers to questions that require calculation can be found on the accompanying CD.

1 A system is oscillating such that the period T is related by the following expression:

$T = kx^n$, where x is the length of the oscillator in mm, and k and n are constants.

The length x is varied and the period T is measured. The table below shows the results.

x/mm	200	250	300	350	400	450	500
T/s	0.622	0.696	0.762	0.823	0.880	0.933	0.984

a Plot a suitable graph that will allow you to determine k and n. [8]
b Use the graph to determine the value of k and n. [4]

2 A resonant tube is such that the length of air inside it can be varied using a movable piston. A loudspeaker is placed at the opened end of the tube. For a given length l, the frequency of the sound coming from the loudspeaker is adjusted so that resonance occurs. Resonance occurs at the fundamental frequency f. The length of the tube is varied and the fundamental frequency is measured. The table below shows the results of the experiment.

Fundamental frequency f/Hz	429	286	215	172	143
Length l/m	0.20	0.30	0.40	0.50	0.60

a Plot a graph of $\frac{1}{f}$ against l. [4]
b Use the graph to determine the speed of sound in the tube. [5]

3 In an experiment to measure the specific heat capacity of a 1.5 kg block of copper, an electric heater is used to heat it and the temperature T at various times is measured. The electric heater is rated at 150 W. The results are shown in the table below.

Time/s	0	90	180	270	360	420
Temp T/°C	26.2	49.2	72.3	95.3	118	134

a Plot a graph of temperature against time. [4]
b Use the graph to determine the specific heat capacity of copper. [5]
c Determine the heat capacity of the block of copper. [1]

4 A long tube is filled with a viscous fluid. Several small metal spheres of different radii are allowed to reach terminal velocity when falling through the viscous fluid. When the spheres are falling with terminal velocity, the time taken for them to travel through a distance of 90.0 cm is measured. The results of the experiment are shown in the table below.

radius r/mm	1.00	1.50	2.00	2.50	3.00
time t/s	60.0	26.7	15.0	9.6	6.7
velocity v/cm s^{-1}					
lg $(v$/cm s$^{-1})$					
lg $(r$/mm$)$					

In this experiment, the terminal velocity v is related to the radius r of the sphere by the following equation:

$v = kr^n$, where k and n are constants.

  a Copy and complete the table.   [5]

  b Plot a suitable graph that will allow you to determine k and n.   [4]

  c Determine the value of k and n.   [3]

5 In an experiment to measure the Young modulus of wood, a metre rule is attached to the edge of a table. A 300 g mass is attached to free end by a piece of string. The length l of the metre rule hanging over the table is recorded. The corresponding vertical distance d is recorded. The string is moved closer to the table. The values of l and d are recorded. The experiment is repeated for various values of l and d.

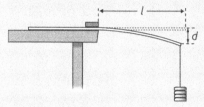

d/m	0.060	0.050	0.032	0.020	0.011	0.005
l/m	0.800	0.750	0.650	0.550	0.450	0.350
l^3/m³						

Width of ruler $w = 3$ cm

Thickness of ruler $t = 6$ mm

It is known that d and l are related by the following equation:

$$d = \left(\frac{4mg}{Ewt^3}\right)l^3$$

where m is the mass attached the metre rule in kg, g is the acceleration due to gravity, E is the Young modulus of wood, w is the width of the ruler in m, and t is the thickness of the ruler in m.

  a Copy and complete the table.   [5]

  b Plot a graph of d against l^3.   [5]

  c Determine the gradient of the line obtained.   [3]

  d Use your graph to estimate the Young modulus of wood.   [3]

6 An alloy in the form of a wire is attached to a rigid support. The original length of the wire is 1.80 m and its diameter is 0.5 mm. Several weights F are attached to the wire and the corresponding extension x is measured. The table below shows the results.

F/N	5	10	15	20	25	30	35	40
x/mm	0.40	0.70	1.10	1.45	1.80	2.30	3.30	5.10

  a Plot a graph of F against x.   [4]

  b Estimate the work done in stretching the wire by 3.0 mm.   [3]

  c Use your graph to determine the Young modulus of the alloy.   [4]

List of physical constants

Universal gravitational constant	G	$=$	$6.67 \times 10^{-11}\,\mathrm{N\,m^2\,kg^{-2}}$
Acceleration due to gravity	g	$=$	$9.80\,\mathrm{m\,s^{-2}}$
Radius of the Earth	R_E	$=$	$6380\,\mathrm{km}$
Mass of the Earth	M_E	$=$	$5.98 \times 10^{24}\,\mathrm{kg}$
Mass of the Moon	M_M	$=$	$7.35 \times 10^{22}\,\mathrm{kg}$
Atmosphere	atm	$=$	$1.00 \times 10^5\,\mathrm{N\,m^{-2}}$
Boltzmann's constant	k	$=$	$1.38 \times 10^{-23}\,\mathrm{J\,K^{-1}}$
Coulomb constant		$=$	$9.00 \times 10^9\,\mathrm{N\,m^2\,C^{-2}}$
Mass of the electron	m_e	$=$	$9.11 \times 10^{-31}\,\mathrm{kg}$
Electron charge	e	$=$	$1.60 \times 10^{-19}\,\mathrm{C}$
Density of water		$=$	$1.00 \times 10^3\,\mathrm{kg\,m^{-3}}$
Resistivity of steel		$=$	$1.98 \times 10^{-7}\,\Omega\,\mathrm{m}$
Resistivity of copper		$=$	$1.80 \times 10^{-8}\,\Omega\,\mathrm{m}$
Thermal conductivity of copper		$=$	$400\,\mathrm{W\,m^{-1}\,K^{-1}}$
Specific heat capacity of aluminium		$=$	$910\,\mathrm{J\,kg^{-1}\,K^{-1}}$
Specific heat capacity of copper		$=$	$387\,\mathrm{J\,kg^{-1}\,K^{-1}}$
Specific heat capacity of water		$=$	$4200\,\mathrm{J\,kg^{-1}\,K^{-1}}$
Specific latent heat of fusion of ice		$=$	$3.34 \times 10^5\,\mathrm{J\,kg^{-1}}$
Specific latent heat of vaporisation of water		$=$	$2.26 \times 10^6\,\mathrm{J\,kg^{-1}}$
Avogadro constant	N_A	$=$	6.02×10^{23} per mole
Speed of light in free space	c	$=$	$3.00 \times 10^8\,\mathrm{m\,s^{-1}}$
Permeability of free space	μ_0	$=$	$4\pi \times 10^{-7}\,\mathrm{H\,m^{-1}}$
Permittivity of free space	ε_0	$=$	$8.85 \times 10^{12}\,\mathrm{F\,m^{-1}}$
The Planck constant	h	$=$	$6.63 \times 10^{-34}\,\mathrm{J\,s}$
Unified atomic mass constant	u	$=$	$1.66 \times 10^{-27}\,\mathrm{kg}$
Rest mass of proton	m_p	$=$	$1.67 \times 10^{-27}\,\mathrm{kg}$
Molar gas constant	R	$=$	$8.31\,\mathrm{J\,K^{-1}\,mol^{-1}}$
Stefan–Boltzmann constant	σ	$=$	$5.67 \times 10^{-8}\,\mathrm{W\,m^{-2}\,K^{-4}}$
Mass of neutron	m_n	$=$	$1.67 \times 10^{-27}\,\mathrm{kg}$

Glossary

A

Absolute refractive index The ratio of the speed of light in a vacuum to the speed of light in the medium.

Absolute zero The lowest temperature possible (0 K or –273.15 °C).

Acceleration The rate of change of velocity.

Accommodation The eye's ability to adjust the shape of its lens in order to produce a focused image on the retina.

Accuracy The measure of the closeness of the measured value to the true value.

Air resistance A force which opposes the motion of an object moving through air.

Amorphous A solid where there is a random arrangement of the atoms throughout the structure.

Amplitude The maximum displacement from the equilibrium position.

Angle of incidence The angle between the incident ray and the normal.

Angle of reflection The angle between the reflected ray and the normal.

Angle of refraction The angle between the refracted ray and the normal.

Angular velocity The rate of change of angular displacement.

Antinode Positions midway between nodes on a stationary wave. The amplitude is maximum at this point.

Archimedes' principle When a body is totally or partially submerged in a fluid, it experiences an upthrust which is equal to the weight of the fluid displaced.

Avogadro constant 6.02×10^{23} per mole

B

Base units Are kilogram, metre, second, kelvin, ampere, mole and candela.

Black body A perfect absorber of thermal radiation.

Boiling A substance absorbs energy and changes state from a liquid to a gas without a change in temperature.

Boyle's law The pressure of a fixed mass of gas is inversely proportional to its volume, provided that the temperature is kept constant.

Brownian motion The haphazard motion of particles.

C

Centre of gravity The point through which all the weight of a body appears to act.

Centripetal force The unbalanced force required to keep an object moving in a circular path.

Charles' law The volume of a fixed mass of gas is directly proportional to its absolute temperature, provided that the pressure is kept constant.

Coherent waves Waves that have the same frequency and hence a constant phase difference between them.

Compression The region on a longitudinal wave where the particles are moving towards each other.

Conduction The process by which thermal energy flows in a solid from a region of higher temperature to a region of lower temperature without the movement of the material itself.

Constructive interference Waves meet in phase (with a phase difference of zero, or a path difference of a whole number of wavelengths) giving a greater resultant displacement.

Convection The process by which thermal energy flows from a region of higher temperature to a region of lower temperature due to the bulk movement of the fluid as a result of a density change.

Couple Two equal and opposite forces whose lines of action do not coincide.

Critical angle The angle of incidence for which the angle of refraction is 90° for a ray travelling from a dense to less optically dense medium.

Critical damping The system comes to rest after one oscillation.

Crystalline solid The atoms, ions or molecules are arranged in a regular pattern that repeats itself within the crystal.

D

Damping The process whereby the amplitude of an oscillating system decreases over time.

Degree Celsius A unit used to measure temperature.

Density The mass per unit volume.

Derived quantities Physical quantities other than the base quantities.

Destructive interference Waves meet out of phase (with a phase difference of 180°, or a path difference of an odd number of half wavelengths) giving a reduced or zero resultant displacement.

Diffraction The spreading out of wavefronts when a wave passes the edge of an object or through a gap.

Diffraction grating A piece of glass or plastic with a large number of equally spaced line drawn.

Displacement This is the distance moved from a fixed point in a stated direction.

Drag force A force which opposes the motion of an object in a fluid.

Ductile material A material that can be easily stretched and formed into a wire.

E

Efficiency The ratio of the useful power output to the power input.

Elastic collision A collision in which kinetic energy is conserved.

Elastic deformation The material returns to its original shape and size when the external force is removed from it.

Elastic potential energy The energy possessed by a body when deformed.

Electric current The rate of flow of charge.

Electric potential energy The energy possessed by a charged body due to its position in an electric field.

Electromagnetic spectrum Electromagnetic waves arranged in order of their wavelengths.

Electromagnetic wave A wave that consists of oscillating electric and magnetic fields at right angles to each other.

Electron diffusion A mechanism by which thermal conduction takes place in metals.

Equipotential A line drawn through points having the same gravitational potential.

Energy The capacity or ability to do work.

Evaporation The process by which a liquid changes into a gas without reaching its boiling point.

F

First law of thermodynamics The change in internal energy of a system is equal to the energy supplied to the system plus the work done on the system.

Fixed point A standard degree of hotness that can easily be reproduced.

Focal length The distance between the optical centre of the lens and the focal point of the lens.

Free body diagram A diagram showing the forces acting on a body.

Frequency The number of oscillations per unit time.

Frequency response The range of frequencies that can be detected by the human ear.

Friction A force which opposes motion.

Fundamental frequency The lowest frequency that a vibrating string or pipe can produce.

G

Geostationary satellite A satellite that has a period of 24 hours and appears to be at the same point above the Earth all the time.

Global warming The increase in temperature of the Earth as a result of the greenhouse effect.

Gravitational field The region around a body where a mass experiences a force.

Gravitational field strength The force acting per unit mass.

Gravitational potential The work done in moving unit mass from infinity to a point.

Gravitational potential energy The energy of a body by virtue of its position in a gravitational field.

Greenhouse effect The process by which heat gets trapped in a greenhouse or in the Earth's atmosphere.

H

Harmonic The nth harmonic is n times the fundamental frequency.

Heat capacity The amount of heat energy required to increase the temperature of a substance by one degree.

Heavy damping The system fails to oscillate.

Hooke's law The extension produced is proportional to force applied, provided that the elastic limit is not exceeded.

I

Ideal gas A gas that obeys the gas laws.

Image distance Distance between the image and the optical centre of the lens.

Impulse The product of a force and time for which it acts on an object.

Inelastic collision A collision in which kinetic energy is not conserved.

Inertia The reluctance of a body to start moving when it is at rest or the reluctance of a body to stop moving when it is in motion.

Intensity The power per unit area.

Internal energy The random distribution of the kinetic and potential energies of the particles that make up the substance.

K

Kelvin The SI unit of temperature.

Kinetic energy Energy possessed by a body by virtue of its motion.

Kinetic theory of gases A theory that assumes that a gas is made up of many small particles moving randomly at high speeds.

L

Latent heat The energy required to change the state of a substance without a change in temperature.

Lattice vibration A mechanism by which thermal conduction takes place in solids without free electrons.

Light damping The amplitude of the oscillation eventually decreases to zero over a period of time.

Linear momentum The product of a body's mass and velocity.

Longitudinal wave A wave in which the oscillations are in the same direction as the energy transfer.

Loudness The subjective sensation of sound that depends on the amplitude of the sound.

Lower fixed point The temperature of pure melting ice at standard atmospheric pressure.

M

Mass The amount of matter contained in a body. It is a measure of a body's inertia.

Mean square speed

$$\langle c^2 \rangle = \frac{c_1^{\,2} + c_2^{\,2} + c_3^{\,2} + \ldots + c_N^{\,2}}{N}$$

Mechanical wave A wave that requires a substance through which to propagate.

Molar heat capacity The amount of thermal energy required to increase the temperature of one mole of a substance by one degree.

Molar heat capacity of a gas at constant pressure The amount of energy required to raise the temperature of one mole of a gas by one degree, when the pressure remains constant.

Molar heat capacity of a gas at constant volume The amount of energy required to raise the temperature of one mole of a gas by one degree when the volume remains constant.

Mole The amount of substance that contains the same number of particles in 12 g of carbon-12.

Moment The product of a force and the perpendicular distance of the line of action of the force from a pivot.

Monomer The unit that repeats itself in a polymer.

N

Natural frequency The frequency with which a system oscillates without applying an external periodic driving force.

Newton The force required to give a mass of 1 kg an acceleration of $1 \, \text{m s}^{-2}$.

Newton's first law A body stays at rest or if moving continues to move with uniform velocity unless acted upon by an external force.

Newton's law of gravitation The force of attraction between two bodies is directly proportional to their masses and inversely proportional to the square of the distance between them.

Newton's second law The rate of change of momentum is proportional to the applied force and takes place in the direction in which the force acts.

Newton's third law If a body A exerts a force on body B, then body B exerts an equal and opposite force on body A.

Node Positions on a stationary wave where the amplitude is zero.

Non-renewable energy Energy that cannot be replenished (fossil fuels).

Normal A line drawn at right angles to a surface.

Normal reaction A force that acts at right angles when one body is in contact with another.

O

Object distance Distance between the object and the optical centre of the lens.

Oscillation An object is moving back and forth about a fixed point.

Overtone The notes other than the fundamental note.

P

Path difference The difference in distance travelled by two waves that interfere.

Period The time taken to complete one cycle or revolution.

Phase A measure of the fraction of the oscillation that has been completed.

Phase difference $\phi = \dfrac{\pi}{\lambda} \times 2\pi$

Physical quantity The property of an object or phenomenon that can be measured with an instrument.

Pitch The subjective sensation of sound that depends on the frequency of the sound.

Plastic deformation The material does not return to its original shape and size when the external force is removed from it.

Polarise To restrict the oscillations of a transverse wave to one plane.

Polymers A material that consists of long chains of molecules.

Potential energy Energy possessed by a body by virtue of its state or position.

Potential difference The work done per unit charge in converting from electrical energy to other forms.

Power The rate at which work is being done.

Precision The measure of the reproducibility of a result.

Pressure The force acting normally per unit area.

Pressure law The pressure of a fixed mass of gas is directly proportional to its absolute temperature, provided that the volume is kept constant.

Principal axis A line that passes through the centre of curvature of a lens such that light is neither reflected nor refracted.

Principle of conservation of energy Energy can neither be created nor destroyed, but can be converted from one form to another.

Principle of conservation of momentum For any system, the total momentum before collision is equal to the momentum after collision,

provided that no external forces act on the system.

Principle of moments For a body that is in equilibrium, the sum of the clockwise moments is equal to the sum of the anticlockwise moments about the same pivot.

Progressive wave A wave in which energy is transferred from one point to another. The wave profile moves.

R

Radian The angle subtended at the centre of a circle by an arc equal in length to the radius of the circle.

Random error An error that occurs as a result of the experimenter and results in an error that is above or below the true value.

Rarefaction The region on a longitudinal wave where the particles are moving away from each other.

Real image An image where the rays actually pass through it.

Refraction The change in direction of a wave when travelling between two media and occurs because of a change in speed of the wave.

Refractive index The ratio $\dfrac{\sin i}{\sin r}$

Relative density The ratio of the density of a substance with the density of water.

Renewable energy Energy derived from natural sources (sunlight, waves, wind, geothermal, etc.).

Resonance The amplitude of an oscillating system is enhanced when the frequency of the driver matches the natural frequency of the system.

Resultant vector The vector sum of the vectors acting on a body.

Root mean square speed

$$\sqrt{\langle c^2 \rangle} = \sqrt{\dfrac{c_1^2 + c_2^2 + c_3^2 + \ldots + c_N^2}{N}}$$

S

Scalar quantity A quantity that has magnitude only.

Simple harmonic motion Period motion in which the acceleration is proportional to the displacement from a fixed point and directed to the fixed point.

SI units The International System of Units.

Snell's law The ratio $\dfrac{\sin i}{\sin r}$ is a constant.

Specific heat capacity The amount of heat energy required to increase the temperature of 1 kg of a substance by one degree.

Specific latent heat of fusion The energy required to convert unit mass (1 kg) of a substance from a solid to liquid without a change in temperature.

Specific latent heat of vaporisation The energy required to convert unit mass (1 kg) of a substance from a liquid to a vapour without a change in temperature.

Speed The rate of change of distance.

Stefan's law The rate at which energy is radiated from a black body is proportional to the fourth power of the temperature of the body expressed in kelvin.

Stationary wave A wave formed when two waves of the same frequency and amplitude are travelling in opposite directions and superimpose.

Strain The extension per unit length.

Stress The force acting normally per unit cross-sectional area.

Superposition When two or more waves arrive at a point, the resultant displacement at that point is the algebraic sum of the individual displacements of each wave.

Systematic error A constant error in one direction.

T

Temperature The degree of hotness of a body.

Terminal velocity The maximum constant velocity reached by an object falling in a fluid.

Thermal conductivity The rate of flow of heat per unit area per unit temperature gradient, when the heat flow is at right angles to the faces of a thin parallel-sided slab of the material, under steady state conditions.

Thermal equilibrium The condition under which two objects in physical contact with each other exchange no heat energy. The two objects are at the same temperature and are said to be in thermal equilibrium.

Thermal radiation The process by which thermal energy is transferred from a region of high temperature to a region of low temperature by electromagnetic waves.

Thermistor A non-linear device with a thermally sensitive resistor.

Thermocouple A device used for measuring temperature.

Thermodynamic scale A temperature scale that is independent of the properties of any particular substance.

Thermoelectric effect The generation of an e.m.f. when two dissimilar metals are joined together.

Thermometric property A physical property that varies continuously with temperature.

Threshold of hearing The minimum intensity of sound that can be detected by the human ear.

Threshold of pain The minimum intensity at which pain is experienced in the ear.

Timbre The subjective sensation of sound that depends on the number of overtones produced with a note.

Torque of a couple The product of one of the forces and distance between the two forces.

Total internal reflection When the angle of incidence within the denser medium is greater than the critical angle, the ray is internally reflected.

Transverse wave A wave in which the oscillations are perpendicular to the direction of energy transfer.

Triple point of water The temperature at which ice, water and water vapour are in thermal equilibrium (273.16 K or 0.01 °C.

U

Upper fixed point The temperature of steam above pure boiling water at standard atmospheric pressure.

Upthrust A force acting vertically upwards on an object placed in a fluid.

V

Vector quantity A quantity that has magnitude and direction.

Vector triangle A triangle showing three forces acting on an object.

Velocity The rate of change of displacement.

Virtual image An image where the rays appear to pass through.

W

Wavelength The distance between two successive crests or troughs.

Weight The force exerted by gravity on an object.

Work The product of a force and the distance moved in the direction of the force.

Y

Young modulus Tensile stress divided by tensile strain.

Index

Headings in **bold** indicate glossary terms.